EasyTerms™
Terminology Guidebook
for Microbiology

Copyright 2010, Ed Creager

This edition of EasyTerms is one in a series of simple-to-use, college-level terminology guidebooks.

Although these guidebooks were originally intended for college students, many high school students will also find them helpful as they prepare for college.

Other topics covered in existing or forthcoming editions:

- Anatomy & Physiology (Human)
- Biochemistry
- Biology
- Botany
- Business Management
- Cell Biology
- Ecology
- Genetics
- Nursing
- Nutrition
- Psychology
- Zoology

EasyTerms can help support your educational advancement and can boost the vocabulary of almost anyone who reads it.

For more information on these and other publications, please visit this site:

ApplecreekBooks.weebly.com

and please note the author's "signature book" entitled,

"The Money-Saving Idea Book: Inside Tips for Starving, Students, Frugal Seniors and Every Financial Survivor."

Nearly all of these books are available at a variety of online retailers.

("The Money-Saving Idea Book" © and ™, Ed Creager, 2009.)

Foreword

This Microbiology edition is a simple-to-use, college-level* terminology guidebook and is part of the EasyTerms reference series. In the book, terms are arranged alphabetically within appropriate topic areas. The complete index makes it easy to find any term and its definition.

* These books can also help high school students prepare so that, before they attend college, they'll already know a considerable amount of the terminology they'll need.

A substantial number of the terms defined here have additional definitions outside the scope of the subject being covered. More general definitions and additional meanings, if sought, are to be found in less specialized publications such as dictionaries and encyclopedias.

Please check this website...

www.ApplecreekBooks.weebly.com

...for more information on other available books.

You can sign up at the website for updates on new books, etc., and you can "opt out" at anytime.

Important Notice:

The resources provided hereby, including websites, books and related materials, are intended to provide accurate information regarding the subject matter. All products and services are provided with the understanding that neither the author nor the publisher is engaged in rendering legal, accounting, or other professional or scholarly advice. If expert assistance is needed, the services of a competent professional should be obtained.

EasyTerms™
Terminology Guidebook

<u>Table of Contents</u>

The terms that follow are divided into the topics shown below. The page number on which the topic begins is given. Within each topic, the terms are arranged alphabetically.

Introduction / Scope and History

1.   alga

A photosynthetic, eukaryotic microorganism.

2.   bacterium

A microorganism lacking a nucleus, usually incapable of making its own food.

3.   control variable

Condition held constant during an experiment.

4.   eukaryotic

Having a distinct nucleus and other membrane-bound structures.

5.   experimental variable

The factor under study, which is allowed to vary, in an experiment.

6.   fungus

Spore bearing eukaryote with absorptive nutrition.

7.   hypothesis

A tentative explanation for a set of observations.

8.   Koch's postulates

Postulates used to prove a certain organism causes a particular disease.

9.   microbiology

The study of microscopic organisms.

10.   microorganism

An organism too small to be seen with the unaided eye.

11. prediction

What one expects to happen in an experiment if a hypothesis is true.

12. prokaryotic

Lacking a membrane bound nucleus.

13. protozoan

A eukaryotic protist with organelles functioning like organs of complex animals.

14. spontaneous generation

The origin of living organisms from nonliving material (a rejected hypothesis).

15. theory

Principles and concepts supported by extensive observation and testing that explain natural phenomena.

16. variable

Condition that can vary during a scientific experiment.

17. virus

An acellular agent containing a nucleic acid and able to replicate only inside a living cell.

Microscopy and Staining

18.   absorption

Process by which light rays are neither transmitted nor reflected from an object.

19.   acid dye

Dye having negative charge.

20.   acid-fast staining

A procedure for identifying acid-fast bacteria.

21.   anionic dye

A positively charged dye.

22.   basic dye

A positively charged dye.

23.   binocular

Having two eye pieces, as in a binocular microscope.

24.   body tube

The part of a microscope that carries an image from the objective to the eyepiece.

25.   bright-field

A kind of microscopy in which subject is illuminated directly so dark object is seen on a bright background.

26.   cationic dye

A dye with a positive ion used for staining microorganisms.

27.   coarse adjustment

Control that changes distance between specimen and lens and brings it nearly into focus.

28.   **compound light microscope**

Microscope that uses ordinary light and has at least two
sets of lenses.

29.   **condenser**

Part of a microscope that concentrates light on the
specimen.

30.   **dark-field**

Viewing of an object under a microscope with light
rflected from the object; light image on dark background.

31.   **differential stain**

Stain used to distinguish between microorganisms.

32.   **diffraction**

Breaking of light waves into bands of different wavelengths.

33.   **electron microscope (EM)**

A microscope that uses an electron beam and electromagnets
instead of light and lenses.

34.   **fine adjustment**

Mechanism on a microscope that alters the distance between
the specimen and ocular very slowly.

35.   **fixation**

Process by which internal parts of cells are preserved in
a particular position while being prepared for study.

36.   **flagellar staining**

Procedure for coating flagella with a dye or metal.

37.   **fluorescence microscope**

A microscope that uses ultraviolet light to excite molecules
so the release light of different colors.

38.   **fluorescent**

The reemission of absorbed light at a different wavelength
(color) than that absorbed.

39.   **fluorescent antibody staining**

Use of fluorescent microscopy and fluorochrome-stained antibodies to locate an antigen.

40.   **fluorochrome**

A fluorescent dye.

41.   **freeze-etching**

Procedure for evaporating water under vacuum from a specimen for electron microscopy.

42.   **freeze-fracturing**

Procedure for freezing and breaking apart cells for study with an electron microscope.

43.   **Gram stain**

A differential stain that distinguishes Gram positive and Gram negative bacteria.

44.   **hanging drop**

A kind of wet mount that allows microscopic observation of motility.

45.   **heat fixation**

Passing a smear of organisms on a slide through an open flame to kill, affix to slide, and render more stainable.

46.   **immersion oil**

Oil with a refractive index equal to glass used between a slide and a microscope lens.

47.   **index of refraction**

Degree to which light rays are bent as they pass from one medium to another.

48.   **iris diaphragm**

Device on a light microscope that regulates the amount of light passing through the specimen.

49.   **light microscopy**

Use of a microscope that focuses any kind of light.

50. **luminescence**

Giving off of absorbed light waves at longer wavelengths.

51. **mechanical stage**

Device attached to a microscope stage that allows precise control of slide movement.

52. **monocular**

Having a single eyepiece, as in a monocular microscope.

53. **mordant**

An agent that helps fix a dye to a cell.

54. **negative staining**

Using dye to create a dark background on which microorganisms can be seen.

55. **numerical aperture**

A property of a lens that indicates how much light can enter it and how great resolution it can achieve.

56. **objective lens**

A microscope lens that creates first image of a specimen.

57. **ocular lens**

A microscope lens that further magnifies an image from an objective lens.

58. **ocular micrometer**

A devise used with an ocular lens to measure actual size of specimens.

59. **optical microscope**

A microscope that uses light to make specimens observable.

60. **parfocal**

Attribute of a microscope in which specimen stays in focus when switched from one objective to another.

61. **phase-contrast microscopy**

Use of light and a condenser to accentuate differences in refractive indexes of structures in a specimen.

62. **reflection**

Light bouncing off an object.

63. **refraction**

Bending of light at an interface between two materials.

64. **refractive index**

Ratio of bending of light in first medium to that in second medium.

65. **resolution**

Ability of an optical device to make two structures identifiable as separate objects.

66. **resolving power**

A numerical expression of the resolution of an optical device.

67. **scanning electron microscope (SEM)**

Microscope that uses electron beams to make surfaces of specimens observable.

68. **shadow casting**

Coating of specimens with metal for observing with an electron microscope.

69. **simple stain**

Use of a single dye to reveal cell shapes and arrangements.

70. **stain**

A dye molecule that binds to structures in organisms and gives them color.

71. **total magnification**

Product of magnifications provided by a microscope's lenses.

72.   **transmission electron microscope (TEM)**

A microscope that uses an electron beam to make internal cell structures observable.

73.   **wavelength**

The distance between successive troughs and crests of a light wave.

74.   **wet mount**

A preparation for microscopic observation in which the specimen in a drop of fluid is placed on a slide.

Fundamentals of Chemistry

75. **acid**

An ionizing substance that donates hydrogen ions.

76. **adenosine diphosphate (ADP)**

The product of an energy expending reaction.

77. **adenosine triphosphate (ATP)**

The body's main energy storage molecule.

78. **alkaline**

Basic, able to accept hydrogen ions.

79. **amino acid**

A molecule having both acid and amino functional groups.

80. **anabolism**

Synthetic metabolic reactions.

81. **anion**

A negatively charge ion.

82. **atom**

Smallest particle that retains properties of an element.

83. **atomic number**

The number of protons in the nucleus of an atom.

84. **atomic weight**

The total number of protons and neutrons in an atom; the average number if there are isotopes of the element.

85. base

An ionizing substance that accepts hydrogen ions or reacts with an acid to form a salt.

86. bioenergetics

The science of energy changes in living systems.

87. buffer

A substance that resists pH change by holding or releasing hydrogen ions in a solution.

88. carbohydrate

An organic compound having several alcohol groups and an aldehyde or ketone group.

89. catalyst

A substance that increases a chemical reaction rate.

90. cation

A positively charged ion.

91. colloid

Glue-like; a particle in a colloidal dispersion.

92. colloidal dispersion

A state of matter with small particles suspended in a medium.

93. complementary base pairing

Combining of adenine with thymine or uracil and cytosine with guanine; basis of information flow in nucleic acids.

94. compound

A substance with two or more elements combined in definite proportion.

95. covalent bond

A chemical bond formed by shared electrons between two atoms.

96.   dehydration

Removal of water.

97.   denaturation

An alteration in the shape and properties of a protein molecule.

98.   deoxyribonuclease

An enzyme that digests DNA.

99.   deoxyribonucleic acid (DNA)

A nucleic acid in chromosomes that directs protein synthesis and transmits genetic information to a new generation.

100.   disaccharide

A molecule having two sugar (saccharide) units held together by a glycosidic bond.

101.   electron

A negatively charged particle that continually moves around the nucleus of an atom.

102.   element

A fundamental unit of matter.

103.   endergonic

Requiring energy, as in a chemical reaction.

104.   entropy

Tendency toward chaos or disorder.

105.   enzyme

A protein that increases the rate of a chemical reaction in a living organism.

106.   exergonic

Releasing energy, as in a chemical reaction.

107. fat

A molecule containing glycerol and one to three fatty acids.

108. fatty acid

A long hydrocarbon chain with a carboxyl group at one end.

109. functional group

A component of a molecule that participates in a chemical reaction.

110. glycine

An amino acid with the simplest chemical structure.

111. glycolipid

A molecule that contains both carbohydrate and lipid components.

112. glycoprotein

A molecule that contains both carbohydrate and protein components.

113. glycosidic bond

A covalent bond between units of a carbohydrate.

114. gram molecular weight (mole)

A quantity of a substance equal to its molecular weight in grams.

115. high-energy bond

A chemical bond that releases more than the unusual amount o energy when the molecule it is in is hydrolyzed.

116. hydrogen bond

Weak covalent bond between hydrogen and another element, such as oxygen or nitrogen.

117. hydrolysis

The splitting of a molecule with the addition of water.

118. **hydrophilic**

Attacted to water.

119. **hydrophobic**

Tending to avoid water.

120. **ion**

A charged atom or group of atoms.

121. **ionic bond**

A chemical bond with atoms held together by the attraction of unlike charges.

122. **isomer**

A molecule having the same kinds and number of atoms as another molecule, but arranged differently.

123. **isotope**

An atom having a different number of neutrons than certain other atoms of the same element.

124. **kinetic**

Energy of motion.

125. **lecithin**

A phospholipid characteristic of animal tissues.

126. **lipid**

Any fat or fatlike substance.

127. **mixture**

Two or more substances combined in any proportions and retaining their individual properties.

128. **mole**

A gram molecular weight.

129. **molecule**

The smallest quantity of a substance that retains its chemical properties.

130. **monosaccharide**

A simple sugar.

131. **neutron**

An uncharged particle in the nucleus of an atom.

132. **nonpolar**

Uncharged; lacking polarity.

133. **nucleic acid**

A polymer of nucleotides; DNA or RNA.

134. **nucleotide**

A molecule having a nitrogenous base, a 5-carbon sugar, and one or more phosphates.

135. **nucleus**

Central part of an atom or a cell; a group of cell bodies in the central nervous system.

136. **organic**

Containing carbon.

137. **oxidation**

Addition of oxygen or loss of electrons in a chemical reaction.

138. **peptide bond**

A chemical bond between the amino group of one amino acid and the carboxyl group of another.

139. **pH**

A scale for expressing acidity or alkalinity; the negative logarithm of the hydrogen ion concentration.

140.   phospholipid

A lipid made of glycerol, fatty acids, and phosphoric acid.

141.   polar compound

A molecule having a charged area or polarity.

142.   polymer

A molecule consisting of repeating units.

143.   polypeptide

A chain of amino acids held together by peptide bonds.

144.   polysaccharide

A molecule consisting of many saccharide units connected by glycosidic bonds.

145.   potential energy

Energy due to position and capable of being released, as in a rock at the top of a hill.

146.   primary structure

The sequence of amino acids in a polypeptide.

147.   product

A substance formed, as in a chemical reaction or process.

148.   protein

A polymer of amino acids.

149.   proton

A positively charged particle in the nucleus of an atom.

150.   quaternary structure

The multiunit three-dimensional structure of a protein.

151. radiation

Spreading from a center; giving off electromagnetic particles and waves.

152. radioisotope

An isotope of an element that gives of radioactive particles and energy.

153. reactant

A substance that is changed by a chemical reaction.

154. reduction

Gain of an electron or hydrogen or loss of oxygen; the realignment of a fractured bone.

155. ribonucleic acid (RNA)

A nucleic acid made from information in DNA that is involved in protein synthesis.

156. rule of octets

Idea that an element is stable if its outer orbit contains eight electrons.

157. saturated fatty acid

A fatty acid lacking double bonds in the carbon chain and being saturated with hydrogen.

158. saturation

Condition of having all chemical affinities satisfied.

159. secondary structure

The bending and coiling of a polypeptide into an helix or other simple structure.

160. secretion

A cell product; the active transport of substances from the blood to the kidney filtrate.

161. solvent

A substance in which other substances can dissolve.

162. specific heat

The amount of heat needed to increase the temperature of a specific volume of substance one degree Celsius.

163. specificity

The attribute of being specific.

164. stereoisomer

Compound having the same kind and number of atoms as another compound, but in a different spatial arrangement.

165. steroid

A lipid with a complex four-ring structure.

166. structural protein

A protein that forms part of a cell structure.

167. template

Pattern.

168. tertiary structure

The three-dimensional globular or fibrous shape of a protein such as results from folding of a helix on itself.

169. trace element

A chemical element normally present in very small amounts in the body.

170. triglyceride

A triacylglycerol (glycerol and three fatty acids).

171. unsaturated fatty acid

Fatty acid with pairs of hydrogen atoms replaced by double bonds in the carbon chain.

172. uridine triphosphate (UTP)

A high energy molecule.

173.   valence

An ion's charge.

Characteristics of Cells

174. active transport

Transport of a substance against a gradient using a carrier molecule, enzyme, and cellular energy.

175. adsorptive endocytosis

Entry of a substance into a cell by first attaching to the cell membrane.

176. amphitrichous

Having a single flagellum at each end.

177. anaphase

A mitotic stage during which chromosomes move apart.

178. axial filament

Structure that allows a spirochete to move.

179. bacillus

Bacterium with a rodlike structure.

180. basal body

Cylindrical structure that attaches a flagellum to a cell.

181. benign

Nonmalignant, favorable for recovery.

182. binding site

A site where a particular molecule binds to a membrane or other structure.

183. bulk flow

Streaming of molecules that allows them to move faster than by diffusion.

184. **cachexia**

Wasting, weakness, and weight-loss seen especially in cancer patients.

185. **capsule**

A secreted protective structure outside a cell wall.

186. **carcinogen**

A cancer-inducing agent.

187. **carrier**

A transfer molecule; a person capable of transmitting an unexpressed gene.

188. **carrier saturation**

A condition with all carrier molecules carrying a substance.

189. **cell**

A basic functional unit of a living organism.

190. **cell cycle**

A repetitive sequence of events in DNA replication and cell division.

191. **cell membrane**

Lipid and protein compounds that form the boundary of a cell.

192. **cell theory**

A theory stating that living things are composed of cells.

193. **cell wall**

Outer layer of certain microorganisms that maintains shape.

194. **centriole**

One of a pair of intracellular bodies that participate in forming a mitotic spindle.

195. **chemoreceptor**

Structure that detects changes in the concentration of a chemical substance.

196. **chemotaxis**

Movement of an organism toward or away from a chemical substance.

197. **chloroplast**

A eukaryotic plastic containing chlorophyll that is the site of photosynthesis.

198. **chromatin**

Nuclear material that condenses into distinct chromosomes during cell division.

199. **chromosome**

Threadlike structure containing DNA.

200. **chrononcology**

The use of cell division rhythms to schedule cancer therapy.

201. **chronotherapy**

The use of any rhythms to schedule therapy.

202. **cilium**

A tiny hairlike projection found on some epithelial cells.

203. **cisterna**

Reservoir or cavity.

204. **coccus**

Spherical bacterium.

205. **cortex**

Laminated heat resistant layer of an endospore.

206. crista

Infolding of inner mitochondrial membrane.

207. cytokinesis

Division of the cytoplasm that follows division of a nucleus.

208. cytoplasm

Cell substance, excluding the nucleus.

209. cytoskeleton

The organelles forming a cell's internal framework.

210. cytosol

The fluid part of cytoplasm that suspends organelles.

211. dipicolinic acid

Acid in endospore cortex that contributes to heat resistance.

212. diplococcus

Two cocci in close association.

213. diploid

Having two of each kind of chromosomes (2N).

214. endocytosis

Movement of particles across a membrane into a cell.

215. endoplasmic reticulum (ER)

A network of membranous vesicles within a cell.

216. endosome

A membranous vesicle formed during endocytosis.

217.   endospore

An extremely resistant, thick walled spore that forms inside certain bacterial cells.

218.   endosymbiont theory

A theory that certain prokaryotic cells were incorporated into eukaryotes as endosymbionts and became organelles.

219.   envelope

In bacteria, all structures outside the cell membrane; in some viruses, a membranous layer around a nucleocapsid.

220.   exocytosis

The movement of particles across a membrane out of a cell.

221.   exoenzyme

Enzyme that is secreted by a cell and that acts outside the cell.

222.   extracellular

Outside a cell.

223.   facilitated diffusion

Diffusion down a gradient on a carrier molecule but not requiring cellular energy.

224.   filtration

Passage of a fluid across a membrane by mechanical pressure.

225.   fimbria

In bacteria, an attachment pilus; in humans a structure that helps guide ova into oviducts.

226.   flagellin

Protein subunits of a flagellum.

227.   flagellum

A movable hairlike process on a cell.

228.   **fluid pinocytosis**

Movement of small quantities of fluid across a membrane into a cell; cell drinking.

229.   **fluid-mosaic model**

A model of molecular arrangements in a cell membrane.

230.   **gas vacuole**

Gas filled body that helps keep cyanobacteria and other aquatic microorganisms afloat.

231.   **gel**

A liquid state in a colloidal dispersion.

232.   **germination**

The process of a spore starting to grow.

233.   **glycocalyx**

Collectively, all polysaccharides found external to a cell wall.

234.   **glycogen**

A polymer of glucose made and stored by animals.

235.   **Golgi apparatus**

Membranous vesicles clustered in cells that complete synthesis of secretions.

236.   **gradient**

The rate of change in the magnitude of concentration, pressure, or other variable.

237.   **granum**

One of a stack of thylakoids in a chloroplast stroma.

238.   **haploid**

Having one of a pair of chromosomes.

239. hydrostatic pressure

Force exerted by a fluid.

240. hyperosmotic

Having higher osmotic pressure .

241. hypertonic

Causing movement of water out of cells.

242. hyposmotic

Having lower osmotic pressure than a reference solution.

243. hypotonic

Causing movement of water into cells.

244. inclusion body

Aggregation of reticulate bodies in chlamydia; aggregation of viruses or viral components in viral infection.

245. integral

Relating to an inseparable component.

246. intermediate filament

Protein filament that contributes to the cytoskeleton of eukaryotic cells.

247. interphase

A cell cycle stage during which the cell is not dividing.

248. intracellular

Within a cell.

249. intrinsic

Entirely within.

250.   isosmotic

Having the same osmotic pressure as a reference solution.

251.   isotonic

Causing no net water movement across a cell membrane.

252.   isozyme

An isomer of an enzyme; one of two or more forms of an enzyme that catalyze the same reaction.

253.   L form

Irregularly shaped bacterium with naturally defective cell wall.

254.   ligand

That which binds to a receptor.

255.   lipopolysaccharide (LPS)

A part of the outer layer of the cell wall in gram-negative bacteria.

256.   lophotrichous

Having clusters of flagella at one or both ends of a cell.

257.   lysis

Bursting and disintegration of a cell.

258.   lysosome

Membrane-bound organelle that contains digestive enzymes.

259.   magnetosome

Magnetic particles in some bacteria that allow them to orient in a magnetic field.

260.   magnetotatic bacterium

Bacterium capable of orienting in a magnetic field.

261. **malignancy**

A tendency to become more virulent; a cancerous growth.

262. **malignant**

Cancerous; tending to become more virulent.

263. **meiosis**

Cell division in which gametes are produce.

264. **mesosome**

An invagination of bacterial plasma membrane possibly associated with chromosome replication.

265. **metachromatic granule**

Phosphate granules that exhibit a variety of colors after staining with a simple stain; volutin.

266. **metaphase**

A mitotic stage during which chromosomes align along the equator of a cell.

267. **metastasis**

The transfer of disease from one organ to another.

268. **microfilament**

A small, hollow protein fiber in cytoplasm that aids in movement or forms part of a cytoskeleton.

269. **microtrabecular lattice**

A network of protein strands in the cytoskeleton of a eukaryotic cell.

270. **microtubule**

A cylindrical organelle that forms part of a cell's mitotic spindle.

271. **mitochondrion**

An organelle that contains enzymes for oxidative and energy-capturing processes.

272.   **mitosis**

Nuclear division that produces two identical nuclei.

273.   **monotrichous**

Having a single flagellum.

274.   **murein**

Peptidoglycan.

275.   **mycelium**

A mass of branching hyphae in fungi and some bacteria.

276.   **nuclear**

Of the nucleus.

277.   **nuclear envelope**

Double membrane around the nucleus of a eukaryotic cell.

278.   **nucleoid**

Nuclear region in a bacterium.

279.   **nucleolus**

A body containing RNA within a nucleus.

280.   **nucleoplasm**

The substance of a nucleus.

281.   **O antigen**

A polysaccharide extending from the outer membrane of some gram-negative bacterial cell walls.

282.   **organelle**

A tiny function unit within a cell.

283. osmosis

Diffusion of water from its own higher to a lower concentration.

284. osmotic pressure

Pressure created by osmosis.

285. outer membrane

Structure located outside the peptidoglycan layer of a gram-negative bacterium.

286. passive transport

A process that moves substances without energy expenditure by the organism.

287. pedicle

Layer of organisms at the air-water interface held together by attachment pili in a broth culture.

288. penicillin

A chemotherapeutic agent made by the mold Penicillium.

289. peptidoglycan

A complex polymer forming the main supporting network of a bacterial cell wall.

290. periplasmic space

A space between the cell membrane and outer membrane in a bacterium.

291. peritrichous

Having evenly distributed flagella.

292. peroxisome

An organelle containing oxidative enzymes.

293. phagocytosis

Engulfment into a vacuole and digestion by a scavenger cell.

294. **plasma membrane**

Membrane forming the boundary of a cell.

295. **plasmid**

Small, circular DNA segment in a cell but not part of a chromosome.

296. **plasmolysis**

Removal of water from a cell by osmosis causing cytoplasm to shrivel.

297. **plastid**

A cytoplasmic organelle containing chlorophyll or other pigments and acts in photosynthesis.

298. **pleomorphic**

Variable in shape.

299. **polar flagellum**

A flagellum found at one end of a long cell.

300. **poly-hydroxybutyrate (PHB)**

A linear polymer used as a carbon reserve by many bacteria.

301. **polyribosome**

Several ribosomes using the same mRNA molecule to make protein.

302. **porin protein**

Protein that forms transport channels across the outer membrane of gram-negative cell walls.

303. **prophase**

The first mitotic stage during which the chromosomes become distinct.

304. **prostheca**

A thin extension of a bacterial cell surrounded by cell membrane and cell wall.

305. protoplasm

Cell substance; literally, first formed.

306. receptor

A specific site with which a specific substance can bind; component of a sense organ.

307. regularly structured layer (RS layer)

A highly ordered protein or glycoprotein on many bacterial surfaces.

308. remission

Abatement of disease symptoms or the period during which the abatement occurs.

309. ribosomal RNA (rRNA)

A nucleic acid that forms part of a ribosome.

310. ribosome

Site for protein synthesis consisting of protein and RNA.

311. run

Movements of a bacterium in a straight line.

312. selectively permeable

A membrane property that allows passage of some substances while preventing passage of others.

313. self-assembly

Spontaneous formation of a complex structure without aid of enzymes.

314. sex pilus

A protein appendage involved in bacterial mating or conjugation wherein organism with pilus donates DNA.

315. slime layer

A loose layer of diffuse material outside a bacterial cell wall.

316. **sodium-potassium pump**

Mechanism that actively moves Na ions out of cells and K ions into them against gradients.

317. **sol**

A liquid state of a colloidal dispersion.

318. **solute**

A dissolved substance.

319. **solution**

A liquid containing dissolved substances.

320. **spheroplast**

Nearly spherical cell from partial removal of cell wall, as in penicillin treatment of gram-negative bacteria.

321. **spirillum**

A rigid, spiral-shaped bacterium.

322. **spirochete**

A flexible, spiral-shaped bacterium with periplasmic flagella.

323. **sporogenesis**

Spore formation; sporulation.

324. **sporulation**

Spore formation.

325. **stroma**

Fluid-filled inner part of a chloroplast; inner part of an organ.

326. **surface tension**

Resistance to rupture by the surface film of a liquid.

327. **surface-to-volume ratio**

The surface area of a structure divided by its volume.

328. **Svedberg unit**

A measure of sedimentation during centrifugation.

329. **teichoic acid**

Glycerol or ribitol polymer joined by phosphates found in cell wall of gram-positive bacteria.

330. **telophase**

The last mitotic stage during which nuclei reform.

331. **teratogen**

An agent that causes defective embryonic development.

332. **thylakoid**

A disklike component of a chloroplast.

333. **tonicity**

The degree to which fluid can move into or out of cells.

334. **triacylglycerol**

A lipid molecule containing glycerol and three fatty acids.

335. **tubulin**

A protein that forms intracellular microtubules.

336. **tumble**

Random turning movement of bacterium that has stopped moving in a straight line; twiddle.

337. **tumor necrosis factor**

A substance that causes degeneration and death of tumor cells.

338. **twiddle**

Tumble of a bacterium.

339. **vibrio**

Bacterium having a comma shape.

340. **volutin granule**

Metachromatic granule; phosphate granule.

Metabolism

341. accessory pigment

Photosynthetic pigment that aids chlorophyll in trapping light energy.

342. acetyl coenzyme A (acetyl CoA)

Molecule that enters the Krebs (citric acid) cycle.

343. activation energy

Energy needed to start a chemical reaction.

344. active site

An enzyme surface region to which a substrate binds; catalytic site.

345. adenine (A)

A nitrogenous base found in nucleic acids that pairs with thymine.

346. aerobic respiration

Metabolic oxidation of organic molecules with transfer of electrons to oxygen.

347. agar

Polysaccharide extracted from red algae.

348. alcoholic fermentation

Fermentation that produces ethanol.

349. allosteric site

The noncatalytic site to which enzyme regulators can bind.

350. amphibolic pathway

Metabolic pathway that can yield energy or substances for synthetic reactions.

351. **anaerobic respiration**

Metabolic reactions that obtain energy by transporting electrons to an acceptor other than oxygen.

352. **anaplerotic reaction**

A chemical reaction that replenishes Krebs cycle intermediates.

353. **anoxygenic photosynthesis**

Photosynthesis that does not split water molecules.

354. **apoenzyme**

The protein portion of an enzyme.

355. **autolysin**

Enzyme that partially digests peptidoglycan so the cell wall can be enlarged.

356. **autotroph**

Organism that uses carbon dioxide as its main carbon source.

357. **auxotroph**

Organism that by losing the ability to synthesize a needed nutrient must obtain it from the environment.

358. **beta oxidation**

A metabolic pathway that breaks fatty acids into 2-carbon segments.

359. **biosynthetic pathway**

A metabolic pathway that makes more complex molecules from simpler ones; anabolism.

360. **butanediol fermentation**

Fermentation that produces butanediol.

361. **Calvin cycle**

Main chemical pathway for carbon fixation in carbon dioxide.

362. carboxysome

Inclusion bodies that contain carbon fixing enzymes.

363. carotenoid

One of several yellowish pigments that aid chlorophyll in capturing light energy.

364. catabolic pathway

A metabolic pathway in which molecules are broken down.

365. catabolism

The breakdown of large molecules into smaller ones.

366. catalytic site

Site at which substrate attaches to an enzyme; active site.

367. chemical equilibrium

Situation in which no net change occurs in concentrations of reactants and products of a chemical reaction.

368. chemiosmosis

Process by which electrochemical gradients account for capture of energy in ATP.

369. chemiosmotic hypothesis

Theory that explains how electron transport and oxidative phosphorylation occur.

370. chemoautotroph (chemolithothrop)

Chemolithotrophic autotroph.

371. chemoheterotroph

Organism that uses organic compounds for energy and a carbon source.

372. chemolithotroph autotroph

Organism that uses inorganic compounds and carbon dioxide.

373.   **chemotroph**

Organism that obtains energy from chemical compounds.

374.   **chlorophyll**

A green pigment containing magnesium that captures energy from light.

375.   **citric acid cycle**

Krebs cycle.

376.   **coenzyme**

Substance that aids an enzyme.

377.   **cofactor**

Nonprotein component of an enzyme.

378.   **colony**

An aggregate of microorganisms descended from a single organism.

379.   **competitive inhibitor**

A substance that inhibits an enzyme by competing with its substrate for occuping the enzyme's active site.

380.   **complex medium**

A culture medium that can vary from batch to batch and that can contain some unknown materials.

381.   **conformational change hypothesis**

Theory that energy from electron transport is used to induce changes in shape of ATP producing enzyme.

382.   **cyclic photophosphorylation**

Reactions by which excited electrons from chlorophyll generate ATP with no water split or carbon dioxide fixed.

383.   **cytochrome**

An electron-carrying heme protein, usually in electron transport chain.

384. **cytosine (C)**

A nitrogenous base found in nucleic acids that pairs with gaunine.

385. **dark reaction**

Component of photosynthesis in which carbon dioxide is reduced by electrons from NADP to form carbohydrates.

386. **deamination**

Removal of an amine group.

387. **defined medium**

Medium having all components present in known quantity.

388. **denitrification**

Process that reduces nitrates to nitrogen gas or nitrous oxide.

389. **differential media**

Culture medium used to distinguish between groups of microorganisms.

390. **electron transport chain**

A set of coenzymes that carry electrons to oxygen.

391. **Embden-Meyerhof pathway**

A biochemical pathway that degrades glucose to pyruvate via fructose-1,6-diphosphate.

392. **endoenzyme**

An enzyme that acts within the cell that made it.

393. **energy**

The capacity to do work.

394. **enthalpy**

The heat content of a system; total energy in a living system.

395. Entner-Doudoroff pathway

A biochemical pathway that degrades glucose to pyruvate and glyceradehyde-3-phosphate.

396. enzyme-substrate complex

An association of an enzyme with its substrate.

397. equilibrium

In a system, a state in which no net change is occurring and free energy is at a minimum.

398. equilibrium constant

A constant characteristic of a chemical reaction at equilibrium determined by concentrations of substances.

399. fatty acid synthetase

A complex of enzymes that synthesize fatty acids.

400. feedback inhibition

Regulation of a metabolic pathway by the product of one reaction inhibiting an enzyme earlier in the pathway.

401. fermentation

Anaerobic metabolism of pyruvate from glycolysis.

402. first law of thermodynamics

A natural law that energy can neither be created nor destroyed, only transformed and redistributed.

403. flavin adenine dinucleotide (FAD)

An electron carrying coenzyme involved in energy production.

404. free energy change

Change in the amount of energy available to do work in a system as system goes from initial to final state.

405. gluconeogenesis

The making of glucose from noncarbohydrate substances.

406. **glycolysis**

The breakdown of glucose to pyruvate.

407. **glycolytic pathway**

Pathway by which glucose is converted to pyruvate; Embden-Meyerhof pathway.

408. **guanine (G)**

Nitrogenous base found in nucleic acids that pairs with cytosine.

409. **heterolactic fermenter**

A microorganism that ferments sugars to lactate and other products such as ethanol and carbon dioxide.

410. **heterotroph**

An organism that uses preformed organic molecules or other organisms as its food source.

411. **hexose monophosphate pathway**

Pentose phosphate pathway.

412. **high-energy molecule**

A molecule containing at least one high energy bond.

413. **holoenzyme**

Apoenzyme and cofactor.

414. **homolactic fermenter**

An organism that ferments sugars almost completely to lactic acid.

415. **Krebs cycle**

A biochemical pathway that degrades acetyl-CoA to carbon dioxide and water with energy capture in ATP.

416. **lactic acid fermentation**

Fermentation that produces mainly lactic acid.

417. **light reaction**

Component of photosynthesis in which light energy is captured.

418. **lithotroph**

Organism that obtains energy from inorganic materials.

419. **macromolecule**

A large molecule that is a polymer of smaller ones.

420. **metabolic pathway**

A set of chemical reactions in which the product of one reaction becomes the reactant in the next.

421. **metabolism**

The sum of all chemical reactions in a living organism.

422. **Michaelis constant**

In enzyme kinetics, a constant equal to the substrate concentration for enzyme to work at half maximal velocity.

423. **mixed acid fermentation**

Fermentation that produces a mixture of organic acids.

424. **mixotrophic**

Combining autotrophic and heterotrophic metabolic processes.

425. **nicotinamide adenine dinucleotide (NAD+)**

An electron carrying coenzyme associated with aerobic metabolism.

426. **nitrification**

Combination with nitrogen.

427. **nitrifying bacterium**

Chemolithotroph that converts ammonia or nitrite to nitrate.

428.   nitrogen fixation

Reduction of nitrogen gas from the atmosphere to ammonia.

429.   nitrogenase

An enzyme that catalyzes nitrogen fixation.

430.   noncompetitive inhibitor

A molecule that inhibits an enzyme by binding to an allosteric site and distorting the enzyme's shape.

431.   noncyclic photophosphorylation

Use of light energy to make ATP by moving electrons from water to NADP.

432.   nucleoside

A molecular unit containing a nitrogenous base and a sugar.

433.   nutrient

Substance an organism needs for growth and reproduction.

434.   organotroph

Pertaining to organism that use organic compounds as their carbon source.

435.   oxidation-reduction reaction

A coupled reaction in which one substance is oxidized and another simultaneously reduced.

436.   oxidative phosphorylation

Synthesis of ATP from ADP using energy made available by electron transport.

437.   oxygenic photosynthesis

Photosynthesis that oxidizes water to oxygen seen in many microorganisms.

438.   passive diffusion

Movement of molecules from regions of higher to lower concentration as a result of random molecular movement.

439. **Pasteur effect**

Decrease in sugar catabolism when an organism switches from anaerobic to aerobic conditions.

440. **pentose phosphate pathway**

A metabolic pathway in which glucose is metabolized to five-carbon sugars and reduced NADP is produced.

441. **peptone**

Water soluble protein hydrolysate used in culture media.

442. **permease**

A membrane carrier protein.

443. **petri dish**

A shallow, round, flat dish used for solid media cultures.

444. **phosphatase**

An enzyme that removes phosphate groups from molecules.

445. **phosphotransferase system**

A system that uses energy from phosphoenolpyruvate to transfer sugars molecules across membranes into cells.

446. **photolithotrophic autotroph**

Organism that uses light energy and inorganic sources of carbon.

447. **photolithotrophic heterotroph**

Organism that uses light energy and simple organic sources of carbon.

448. **photolysis**

Use of energy from excited electrons to split water molecules into oxygen, protons, and electrons.

449. **photosynthesis**

Conversion of light energy to chemical energy then used to reduce carbon.

450. **photosystem I**

System in eukaryotic cells using wavelengths longer than 680 nm in both cyclic and noncyclic photophosphorylation.

451. **photosystem II**

System in eukaryotic cells using wavelengths shorter than 680 nm in noncyclic photophosphorylation.

452. **phototroph**

An organism that uses light as an energy source.

453. **phycobiliproteins**

Photosynthetic pigments consisting of tetrapyrolle and protein found in cyanobacteria and red algae.

454. **phycocyanin**

A blue phycobiliprotein pigment used in photosynthesis.

455. **phycoerythrin**

A red phycobiliprotein pigment used in photosynthesis.

456. **pour plate**

A culture in which microorganisms are added to a liquid medium which is poured into a plate and allowed to harden.

457. **prosthetic group**

A tightly bound component of an enzyme that remains at the active site during catalytic activity.

458. **protease**

An enzyme that breaks proteins into individual amino acids.

459. **protonmotive force (PMF)**

Force from a proton gradient thought to power ATP synthesis and certain other processes.

460. **prototroph**

Microorganism that has nutrient requirements typical of naturally occuring members of the same species.

461. **pure culture**

A culture containing a single kind of organism often derived from a single organism.

462. **purine**

A nitrogenous base having a double ring structure.

463. **pyrimidine**

A nitrogenous base having single ring structures.

464. **ribulose-1,5-bisphosphate carboxylase**

Enzyme that catalyzes carbon fixation in the Calvin cycle.

465. **second law of thermodynamics**

A law stating that physical and chemical processes tend to gain entropy and lose organization.

466. **selective medium**

Culture medium with ingredients that select for the growth of certain organisms at the expense of others.

467. **siderophore**

A small molecule that complexes with iron thereby aiding in its transport into cells.

468. **spread plate**

A solid culture in a petri dish made by spreading dilute organisms over agar surface for isolating colonies.

469. **streak plate**

Culture on solid medium in a petri dish with colonies distributed for isolation.

470. **substrate**

Substance acted on by an enzyme.

471. **substrate-level phosphorylation**

Making of ATP from ADP at the same time another energy containing molecule is broken down.

472. **synthetic medium**

A chemically defined medium.

473. **thermodynamics**

A science that deals with energy and its interconversions.

474. **thymine (T)**

A nitrogenous base found in nucleic acids that pairs with adenine.

475. **transamination**

Exchange of an amino group between two molecules.

476. **tricarboxylic acid (TCA) cycle**

Krebs cycle, citric acid cycle.

477. **uracil (U)**

A nitrogenous base found in RNA that pairs with adenine.

478. **vitamin**

An organic nutrient required in small quantities by an organism that cannot synthesize it.

Growth and Culturing

479. **acidophile**

A microorganism with optimum growth below pH 5.5.

480. **activation**

Initiation of germination by a traumatic agent.

481. **aerobe**

An organism that can grow in the presence of atmospheric oxygen.

482. **aerotolerant anaerobe**

Microorganisms that do not use oxygen but can grow in its presence.

483. **agar plate**

A plate of medium solidified with agar.

484. **alkalinophile**

A microorganism with optimum growth between pH 8.5 and 11.5.

485. **anaerobe**

Organism that grows in the absence of oxygen.

486. **aseptic technique**

One of several procedures used to minimize contamination of microbial cultures with organisms from the environment.

487. **axial nucleus**

Long nucleus formed during sporulation.

488. **barophilic**

Organisms that grow and reproduce best under high pressure.

489. barotolerant

Organisms that can grow and reproduce under high pressure.

490. batch culture

A culture of organisms made by inoculating a closed culture vessel with a single batch of medium.

491. binary fission

Asexual separation of an organism into two organisms.

492. blood agar

A medium containing blood and solidified with agar.

493. budding

Asexual reproduction in which small new organism separates from larger parent.

494. casein hydrolysate

A milk protein product that contains amino acids and is used to enrich media.

495. chemostat

A continuous culture apparatus that replenishes medium as it is used.

496. chocolate agar

Agar that contains blood and has been heated turning blood brown.

497. coenocytic

Having multinucleate cells resulting from nuclear divisions without cytoplasmic divisions.

498. colony forming unit (CFU)

The number of organisms that can form colonies in a plate; an indication of the number of viable organisms present.

499. continuous culture system

System with constant environmental conditions maintained by renewing nutrients and removing wastes.

500. core

Living portion of an endospore.

501. dark reactivation

Excision and replacement of thymine dimers in DNA in absence of light.

502. daughter cell

One of progeny from cell division.

503. death phase

Decline phase in a microbial growth curve.

504. decline phase

Phase during which cells lose ability to divide.

505. defined synthetic medium

Manufactured medium with all components present in known quantity.

506. direct microscopic count

Determining the number of cells in a culture by counting a sample in a calibrated counting chamber.

507. doubling time

Time for organisms in a culture to double in number.

508. enrichment medium

A culture medium that contains special nutrients to support growth of a particular organism.

509. eurythermal

Pertaining to organisms that grow well over a wide temperature range.

510. exponential growth rate

Growth (cell division) characterized by doubling of the number of organisms in a fixed time interval.

511. exponential phase

The phase of a growth curve during which growth is exponential, or logarithmic.

512. extracellular enzyme

An enzyme that acts outside a cell.

513. facultative

Able to tolerate the presence or absence of an environmental condition.

514. facultative anaerobe

Microorganism that uses oxygen when present but that shifts to anaerobic metabolism when oxygen is not available.

515. facultative psychrophiles

Organism that grows best below 20 degrees C but that can grow above that temperature.

516. facultative thermophile

Organism that grows both above and below 37 degrees C.

517. fastidious

Having special nutritional needs that make an organism difficult to grow in the laboratory.

518. generation time

Time needed for a microbial population to double in number.

519. growth

Among microbes, increase in numbers; in most organisms, increase in body size.

520. growth yield (Y)

A quantitative measure of mass produced by supplying a nutrient.

521. halophile

An organism that requires moderate to large salt concentrations.

522. involution

Formation of unusual cell shapes during the decline phase of bacterial growth.

523. ionizing radiation

High energy radiation that causes atoms to lose electrons (ionize).

524. lag phase

Portion of growth curve after microorganisms are introduced into fresh medium before they increase in number.

525. log phase

Portion of a growth curve during which cells divide at an exponential, or logarithmic rate.

526. mean growth rate constant (k)

Number of generations per unit time in a microbial population.

527. membrane filter

A thin filter with carefully control pore size used to remove microorganisms from solutions.

528. mesophile

A microorganism that grows best in the range of 20 to 45 degrees C.

529. microaerophile

A microorganism that requires a low oxygen concentration of 2 to 10 percent.

530. microbial growth

Increase in numbers of organisms or cells.

531. most probable number (MPN)

A statistical estimate of microbial growth using dilution methods when there are too few organisms for plate counts.

532. mother cell

Cell that has approximately doubled in size and is ready to divide.

533. **neutrophile**

A microorganism that grows best in a pH range of 5.5 to 8.0.

534. **nonsynchronous growth**

Growth of microorganisms in which all cells divide during generation time but not at same time.

535. **nutritional factor**

A nutrient or other agent that influences what organisms are present in an environment and their growth.

536. **obligate**

Required.

537. **obligate aerobe**

An organism that requires oxygen.

538. **obligate anaerobe**

An organism that requires an environment without free oxygen.

539. **obligate psychrophile**

An organism that grows only below 20 degrees C.

540. **obligate thermophile**

An organism that grows only above 37 degrees C.

541. **optimum pH**

The pH at which a microorganism grows best.

542. **osmotolerant**

Able to grow over a wide range of osomotic concentrations.

543. **outgrowth**

The final stage in germination in which DNA, RNA, and proteins are synthesized.

544. periplasmic enzyme

An enzyme secreted into and acting in the periplasmic space.

545. photoreactivation

Use of blue light and photoreactive enzyme to repair thymine dimers in DNA.

546. physical factor

Environmental factor that influences the kinds of organisms found in an environment and their growth.

547. preserved culture

A culture containing organisms in a dormant state.

548. psychrophile

Organism that grows best below 15 degrees C.

549. psychrotroph

Organism that grows best between 20 and 30 degrees C, but that can grow at 0 degrees C.

550. reference culture

Preserved culture used to maintain organisms with originally defined characteristics.

551. serial dilution

A set of tubes of media containing various logarithmic concentrations of microorganisms.

552. serum

The fluid part of the blood after clotting factors have been removed.

553. stationary phase

Phase in microbial growth in which growth ceases and number in a population level off.

554. stenothermal

Pertaining to an organism that grows well over a limited temperature range.

555. **stock culture**

Pure culture of an isolated organism reserved to replenish a laboratory supply.

556. **strict anaerobe**

Organisms that cannot survive in free oxygen; obligate anaerobe.

557. **synchronous growth**

Hypothetical stair-step growth curve in which all cells divide at the same time.

558. **thermophile**

An organism that grows best above 45 degrees C.

559. **turbidity**

A cloudy appearance in a culture tube indicative of growing organisms.

560. **turbidostat**

A continuous culture system with a photocell that controls media flow to maintain constant turbidity.

561. **ultraviolet light (UV)**

Short wavelength, relatively higher energy radiation.

562. **yeast extract**

Vitamins and other nutrients from yeast used to enrich media.

Microbial Genetics

563.   acridine derivative

Chemical mutagen that acts by insertion between DNA bases.

564.   alkylating agent

Chemical mutagen that adds alkyl groups to DNA bases.

565.   alleles

Genes occupying the same locus but carrying different information.

566.   allosteric enzyme

Enzyme whose activity can be altered by a substance binding to a site other than its active catalytic site.

567.   Ames test

Test used to determine whether a substance can cause mutations.

568.   aminoacyl or acceptor site (A site)

The site on a ribosome where new amino acids are added to a growing polypeptide chain.

569.   antiparallel

Parallel but running in opposite directions.

570.   attenuation

Weakening of a disease-producing organism; termination of transcription of an operon by a genetic control mechanism.

571.   attenuator

The site of genetic control of attenuation.

572.   bacteriocin

Protein released by some bacteria that inhibits other strains of the same or closely related species.

573. **bacteriocinogen**

A plasmid that directs synthesis of a bacteriocin.

574. **bacteriophage**

Any of several viruses that infect a specific bacteria.

575. **base analog**

A molecule that can substitute for a DNA nucleotide during replication thereby causing a mutation.

576. **catabolite repression**

Inhibition of synthesis of enzymes for breaking down one nutrient by the presence of another nutrient.

577. **chimera**

Recombinant plasmid containing foreign DNA used in genetic engineering.

578. **chromosome mapping**

The practice of indentifying the sequence of genes within a chromosome.

579. **clone**

A group of identical cells derived from a single cell.

580. **codon**

A three-base sequence in messenger RNA derived from DNA and specifying amino acid placement in a protein.

581. **colicin**

Protein encoded by plasmid in an enteric bacterium that binds to and lyses particular target bacteria.

582. **competence factor**

A protein released by a cell that facilitates uptake of DNA.

583. **competent**

Having the ability to take up DNA fragments and incorporate them into the genome during transformation.

584. **complementary DNA (cDNA)**

A DNA copy of an RNA molecule.

585. **conditional mutation**

Mutation expressed only under certain environmental conditions.

586. **conjugation**

A kind of sexual reproduction in some protozoa; a kind of gene transfer in some bacteria.

587. **conjugative plasmid**

A plasmid with genes for sex pili active in bacteria during conjugation.

588. **constitutive enzyme**

Enzyme present at all times regardless of nutrients in medium.

589. **constitutive mutant**

Organism with mutation in DNA that produces a constitutive enzyme.

590. **corepressor**

A molecule that inhibits synthesis of repressible enzyme.

591. **crossing-over**

The exchange of genetic information between segments of homologous chromosomes.

592. **deletion**

Loss of one or more bases from a DNA strand.

593. **diauxic growth**

Growth by using one nutrient until depleted and shifting to a second nutrient.

594. **dimer**

Two pyrimidines bonded together in a DNA strand.

595. **DNA ligase**

Enzyme that joins DNA fragments.

596. **DNA polymerase**

An enzyme that increases chain length in DNA synthesis.

597. **DNA replication**

Synthesis of new DNA according to information in an existing DNA template.

598. **electrophoresis**

A separation technique that uses differential migration rates of substances in an electrical field.

599. **elongation cycle**

The part of protein synthesis in which amino acids are added to a growing polypeptide chain.

600. **end product inhibition**

Feedback inhibition; condition in which the product of a set of reactions inhibits an early acting enzyme.

601. **enzyme induction**

Mechanism by which genes for enzymes to metabolize a nutrient are turned on by the presence of the nutrient.

602. **enzyme repression**

Mechanism by which genes for enzymes to make a paraticular metabolite are turned off by the metabolite.

603. **episome**

A plasmid that can exist in or out of a host cells' chromosome.

604. **exogenote**

A segment of donor DNA that enters a bacterial cell during recombination.

605. **exon**

A region in a gene that codes for RNA that appears in the final mRNA.

606. **exonuclease**

An enzyme that removes DNA segments.

607. **expression vector**

A cloning vector used to exprerss recombinant genes in a host; a recombinant gene and the protein it synthesizes.

608. **F factor**

Fertility factor; a plasmid that makes E. coli host the gene donor for bacterial conjugation.

609. **F pilus**

A bridge from F plus to F minus cells during conjugation.

610. **F plasmid**

Extrachromosomal DNA in F plus cells.

611. **F' plasmid**

An F plasmid removed from and carrying part of a bacterial chromosome.

612. **fluctuation test**

Test that demonstrates whether resistance to chemical substances is spontaneous or induced by an outside agent.

613. **frameshift mutation**

A DNA sequence change caused by adding or deleting bases.

614. **gene**

DNA nucleotide sequence that forms a functional unit of a chromosome.

615. **gene amplification**

Inducing plasmids or bacteriophages to reproduce rapidly inside cells by genetic engineering techniques.

616. **general recombination**

Recombination involving reciprocal exchanges of homologous DNA sequences.

617. **generalized transduction**

Transfer of part of a bacterial genome in which the DNA fragment is inadvertently packaged within a phage capsid.

618. **genetic code**

The three-base sequences in messenger RNA derived from a DNA template that determine amino acid order in proteins.

619. **genetic engineering**

Deliberate modification of genetic information by changing an organism's genome.

620. **genetic fusion**

In genetic engineering, transposition of location or coupling of genes from two operons.

621. **genotype**

The genetic information stored within an organism's DNA.

622. **helicase**

An enzyme that uses energy to unwind DNA prior to transcription.

623. **heredity**

The passing on of genetic traits from an organism to its offspring.

624. **heteroduplex DNA**

Double stranded DNA formed by two not completely complementary strands.

625. **heterogeneous nuclear RNA (hnRNA)**

A group of large, variable size pieces of RNA in a nucleus.

626. **Hfr strain**

A bacterial strain that donates genes with high frequency to recipient cells during conjugation.

627. **high frequency of combination strain**

Bacterial strain that frequently donates genes to recipient cell during conjugation because of an F factor.

628.   histone

A small basic protein found in eukaryotic DNA.

629.   host restriction

Breakdown of foreign genetic material that has entered a host cell.

630.   hybridoma

A hybrid cell from fusing a cancer cell with another kind of cell.

631.   induced mutation

Change in DNA caused by an outside agent.

632.   inducer

Molecule that stimulates sunthesis of an inducible enzyme, such as the substrate for the enzyme.

633.   inducible enzyme

Enzyme that varies in concentration depending on what stimulation by an inducer has occurred.

634.   initiating segment

A portion of an F plasmid that is transferred during conjugation of an Hfr bacterium.

635.   insertion

Addition of one or more bases to a DNA, causing a frameshift mutation.

636.   insertion sequence

A transposon containing genes for enzymes required for transposition.

637.   intercalating agent

A molecule that can be inserted between DNA bases distorting the molecule and causing a mutation.

638.   intron

A segment of DNA between coding regions of a gene that is transcribed but later removed from the mRNA product.

639. **leader sequence**

A nontranslated sequence of mRNA that aids in initiation and regulation of transcription.

640. **ligase**

Enzyme that cuts molecules apart.

641. **light repair**

Repair of dimers in DNA by an enzyme activated by light.

642. **locus**

Location of a gene on a chromosome.

643. **lysogenic**

Containing prophages.

644. **lysogeny**

Persistence of prophages in a bacterium with neither virus replication nor cell destruction.

645. **messenger RNA (mRNA)**

A nucleic acid that carries information as codons for protein synthesis.

646. **metabolic channeling**

Location of enzymes and metabolites in particular sites in cells.

647. **missense mutation**

A single base substitution in DNA that changes a codon so it codes for a different amino acid.

648. **monoclonal antibody**

Antibody made in the laboratory by a clone of cultured cells usually from fusing a cancer cell and a lymphocyte.

649. **mutagen**

An agent that can alter DNA.

650.   **mutation**

A change in DNA.

651.   **nonsense codon**

Codon that does not code for an amino acid; termination codon.

652.   **nonsense mutation**

A mutation that changes a sense codon to a nonsense codon.

653.   **nucleosome**

A complex of DNA and histone protein in chromatin of a eukaryotic cell.

654.   **Okazaki fragment**

Short polynucleotide made during discontinuous DNA replication.

655.   **operator**

A DNA segment to which a repressor protein binds; controller of expression of adjacent genes.

656.   **operon**

A set of structural genes and the operator that controls their expression.

657.   **pacemaker enzyme**

Enzyme in a metabolic pathway that catalyzes the slowest, or rate-limiting, reaction.

658.   **peptidyl or donor site (P site)**

A site on a ribosome at which elongation of a polypeptide chain begins.

659.   **peptidyl transferase**

An enzyme that catalyzes elongation of a polypeptide chain.

660.   **phage**

Bacteriophage.

661.   phenotype

Specific, observable characteristics of an organism.

662.   point mutation

A change in DNA that affects one codon.

663.   posttranscriptional modification

Alteration of a protein after its amino acid sequence is established.

664.   Pribnow box

A DNA sequence at the start of transcription in prokaryotic genes to which RNA polymerase binds; TATAAT.

665.   promoter

Region of DNA at the start of a gene where RNA polymerase binds before transcription begins.

666.   prophage

A latent temperate phage remaining within a lysogen and usually integrated into the host chromosome.

667.   protoplast

A cell that has had its wall removed.

668.   protoplast fusion

Joining of cells that have had their walls removed.

669.   recombinant DNA

DNA containing genetic information from two different species.

670.   recombination

Making of a new chromosome by combining DNA from two organisms.

671.   recombination repair

Repair of DNA using a DNA segment from a sister molecule.

672. **regulator gene**

Gene that controls expression of structural genes.

673. **replica plating**

Technique for plating and locating specific mutants.

674. **replicate**

Make an exact copy of.

675. **replication**

Duplication.

676. **replication fork**

Y-shaped structure at site of DNA replication.

677. **replicon**

Unit of a genome that can initiate replication and in which DNA is replicated.

678. **repressible enzyme**

Enzyme whose concentration decreases in the presence of a small molecule, usually a metabolic end product.

679. **repressor protein**

Protein coded by a regulator gene that can bind to an operator and prevent transcription.

680. **resistance gene**

Component of a resistance plasmid that confers resistance to a particular antibiotic.

681. **resistance transfer factor (RTF)**

Component of a resistance plasmid that facilitates plasmid transfer by conjugation.

682. **restricted transduction**

Transduction in which a particular set of bacterial genes are carried to another bacterium by a temperate phage.

683. **restriction endonuclease**

Enzyme that cuts DNA at specific sites in base sequences.

684. **rho factor**

A protein that helps remove RNA polymerase after it has completed transcription.

685. **RNA polymerase**

Enzyme that binds to an exposed DNA strand during transcription.

686. **rolling circle**

A mode of DNA replication that moves around a circular DNA.

687. **semiconservative replication**

DNA replication in which each new molecule consists of one old molecule and one new one made from the old template.

688. **sense strand**

DNA that RNA polymerase copies to make RNA.

689. **sigma factor**

Protein that aids RNA polymerase to recognize the promoter at the start of a gene.

690. **silent mutation**

A change in DNA that does not change the amino acids in a protein.

691. **site-specific recombination**

Recombination of nonhomologous genetic material with a chromosome at a particular site.

692. **specialized transduction**

Transfer of a specific set of bacterial genes by a temperate phage.

693. **spontaneous mutation**

Changes in DNA in the absence of any known mutagenic agent.

694. **structural gene**

A gene that codes for synthesis of a particular protein.

695. **suppressor mutation**

A mutation that counteracts an earlier mutation and restores the normal phenotype.

696. **temperate phage**

Bacteriophage capabale of establishing a lysogenic relation-ship with a bacterium.

697. **terminator**

A codon that marks the end of a gene.

698. **Ti plasmid**

A plasmid from Agrobacterium tumifaciens used to insert genes into plant cells.

699. **transcription**

The transfer of coded genetic information from DNA to mRNA.

700. **transduction**

Transfer of DNA from one bacterium to another by a bacterio-phage.

701. **transfer RNA (tRNA)**

RNA that carries amino acids to specific sites in a growing peptide chain.

702. **transformation**

A change in an organism resullting from the transfer of naked DNA.

703. **transition mutation**

Substitution of one purine for another or of one pyrimidine for another in DNA.

704. **translation**

The process by which mRNA codons are used to determine the sequence of amino acids in a protein.

705.   transposable element

Transposon.

706.   transposition

Movement of a piece of DNA from one site to another in a chromosome.

707.   transposon

A DNA segment with genes for transposition that itself moves about a chromosome.

708.   transversion mutation

Substitution of a purine for a pyrimidine or a pyrimidine for a purine in DNA.

709.   virulent bacteriophage

A bacteriophage that lyses its host cell during its reproductive cycle.

Microbes and Taxonomy

710. archaeobacterium

Primitive prokaryotes lacking peptidoglycan in cell walls.

711. binomial system

A taxonomic system created by Linnaeus in which an organism has a genus and species name.

712. classification

Ordering into groups in some rational way, as by similarity or evolutionary relationship.

713. dendrogram

Branching diagram to show relationships among organisms.

714. dichotomous key

Taxonomic key consisting of paired either-or statements.

715. divergent evolution

Changes that cause descendents of a common ancestor to become less and less alike.

716. DNA hybridization

Process whereby DNA strands from different organisms are combined into a single DNA molecule.

717. eubacterium

A true bacterium.

718. genetic homology

Degree of similarity of DNA among organisms.

719. genus

A major subdivision of living things that includes one or more species.

720. Jaccard coefficient

In numerical taxonomy the proportion of matching characters among those present in two organisms being compared.

721. melting temperature of DNA

Temperature at which paired strands of DNA separate.

722. monera

A taxonomic group containing organisms with prokaryotic cells.

723. natural classification

Classsification based on shared biological characteristics.

724. nomenclature

The naming of organisms and their assignment to taxonomic groups.

725. nucleic acid hybridization

Forming double stranded DNA in which the two strands come from different species.

726. numerical taxonomy

Use of numerical methods to assign organisms to categories based on which characteristics they display.

727. phenetic system

A classification system that groups organisms according to their similarity.

728. phylogenetic (phyletic) system

A classification system based on evolutionary relationships.

729. protein profile

A representation of the proteins a cell contains based on results, such as polyacrylamide gel electrophoresis.

730. protista

A large kingdom of eukaryotic microorganisms.

731. **species**

The narrowest taxon of organisms.

732. **specific epithet**

The second name of an organism in binomial nomenclature.

733. **strain**

Population of descendants of a single cell; pure culture.

734. **stromatolite**

Fossilized prokaryotes deposited in mats.

735. **systematics**

Study of characteristics of organisms to arrange them in an orderly fashion; taxonomy.

736. **taxon**

A group of related organisms.

737. **taxonomy**

Study of characteristics of organisms to arrange them in an orderly, fashion; systematics.

738. **zygote**

A cell resulting from the union of male and female gametes.

Bacteria

739. **acid-fast**

Property of bacteria that cannot be decolorized easily with acid alcohol after staining with basic dye.

740. **Actinomyces**

Filamentous, nonsporing, irregular gram-positive rods once thought to be fungi.

741. **akinete**

Dormant, thick walled cell formed by some fungi and cyanobacteria.

742. **alpha hemolysis**

Incomplete hemolysis of erythrocytes by bacterial enzymes leaving a greenish zone around a colony on blood agar.

743. **beta hemolysis**

Complete hemolysis of erythrocytes by bacterial enzymes leaving a clear area around a colony grown on blood agar.

744. **bioluminescence**

Light production by living cells.

745. **chlamydia**

A tiny, spherical, nonmotile bacterium capable of dividing only inside a cell.

746. **coagulase**

An enzyme that induces blood clotting and is typically produced by pathogenic staphylococci.

747. **conidium**

A conidiospore.

748. **corynebacterium**

Club-shaped, irregular gram-positive rod that does not form spores.

749. cyanobacterium

Prokaryotic, photosynthetic usually unicellular organism.

750. elementary body

A small, dormant transmissible form of chlamydiae.

751. enteric bacterium

Bacteria of the family Enterobacteriaceae, many of which live in the intestinal tract, some causing disease.

752. fruiting body

Structure that holds spores (sexual or asexual) in fungi and some bacteria.

753. heterocyst

Specialized nitrogen fixing cell in cyanobacteria.

754. holdfast

A structure by which some bacteria and algae are fastened to a solid object.

755. hormongonia

Small motile fragments of cyanobacteria active in dispersal and reproduction.

756. initial body

Reticulate body of chlamydia.

757. lactobacillus

Gram-positive rod used in making cheeses and various fermented foods.

758. methanogenic bacterium

An anaerobic bacterium that produces methane.

759. methylotroph

A bacterium that uses methane or methanol as its carbon source.

760. **micrococcus**

A small spherical aerobe or facultative anaerobe that forms regular clusters bu dividing in two or more planes.

761. **mineralization process**

The breakdown of organic to inorganic substances by microorganisms.

762. **mycolic acid**

Very large (60-90 carbon) fatty acids in cell walls of mycobacteria.

763. **mycoplasma**

Small bacteria lacking cell walls and peptidoglycan usually requiring steroids for growth.

764. **myxobaterium**

Gliding, gram-negative soil bacteria.

765. **myxospore**

Dormant spores made by myxobacteria.

766. **nocardioform**

Gram-positive, nonmotile somtimes filamentous and acid-fast organisms, some causing skin and respiratory diseases.

767. **outer sheath**

A multi-layered membrane around the protoplasmic cylinder in spirochetes.

768. **pasteurella-haemophilus group**

A group of small, gram-negative, nutritionally fastidious coccobacilli.

769. **propionobacterium**

Irregular, pleomorphic, gram-positive rods.

770. **protoplasmic cylinder**

Sheath surrounded coiled cylinder of a spirochete.

771. pseudomonads

Aerobic, motile bacteria with polar flagella.

772. pseudomurein

A modified peptoglycan found in methanogenic bacteria.

773. purple membrane

A membrane containing bacteriorhodopsin that captures light energy in halobacteria.

774. reticulate body

An intracellular form of a chlamydia.

775. rickettsia

Small, gram-negative, nonmotile obligate intracellular parasite.

776. sheath

A hollow tube covering a chain of organisms seen in a few groups of bacteria.

777. stalk

A nonliving appendage made by and supporting certain bacteria.

778. streptococcus

Aerotolerant anaerobic cocci most of which lack catalase.

779. streptomycetes

Filamentous, gram-positive spore forming soil bacteria, many of which produce antibiotics.

780. thermoacidophile

An organism that grows best at high temperature and acidic conditions.

781. treponeme

Spirochete of the genus treponema; one is causative agent of syphilis.

782.   type strain

First defined reference strain of a bacterial species.

Viruses

783. accumulation period

Period during which viral components accumulate and are assembled in a cell.

784. acute infection

Disease with rapid onset and relatively short course.

785. adsorption

Attachment of a substance to a surface, such as that of a virus to a host cell.

786. binal symmetry

Symmetry of a virus capsid that combines icosahedral and helical shapes.

787. burst size

Number of phage particles released by a host cell during a lytic cycle.

788. burst time

Time from absorption of phages to release of newly replicated phages.

789. cancer

Disease characterized by malignant cell division.

790. capsid

Protein shell around a virus' nucleic acid.

791. capsomere

A unit of a virus' protein coat.

792. complex virus

Virus with an envelope or other specialized structure such as a head or tail.

793. **concatamer**

A chain of repeating subunits.

794. **cytopathic effect**

Observable change in cells due to viral replication in them.

795. **early mRNA**

Messenger RNA produced early in a viral infection that allows the virus to take over host cell.

796. **eclipse period**

Initial period in viral infection when host bacteria contain no complete viruses.

797. **enveloped virus**

A virus with an envelope.

798. **host range**

Different types of organisms a microbe can infect.

799. **human immunodeficiency virus (HIV)**

Acquired immune deficiency syndrome.

800. **icosahedral**

A three-dimensional shape with twenty equilateral triangular faces.

801. **late mRNA**

Messenger RNA made late in a virus infection and that codes for capsid proteins and proteins needed for virus release.

802. **latency**

A virus' ability to remain in host cells for a long time without losing its ability to replicate.

803. **latent period**

Period before microorganisms increase in number; lag phase.

804. latent virus infection

Infection in which a virus becomes dormant but has the capacity to reactivate.

805. lysogen

A bacterium carrying a viral prophage and having the potential to produce bacteriophages.

806. lytic cycle

A virus life cycle that includes host cell lysis.

807. matrix protein

Protein inside viral envelope that assists in viral assembly.

808. maturation

Process whereby complete virions are assembled from components.

809. minus (negative) strand

A viral nucleic acid strand complementary in base sequence to the viral mRNA.

810. naked virus

A virus lacking an envelope.

811. nucleocapsid

Nucleic acid and protein coat of a virus; basic unit of a virus.

812. obligate intracellular parasite

An organism that can live and reproduce only inside a living host cell.

813. oncogene

A gene which contributes to converting normal cells to cancer cells.

814. penetration

Entry of a virus into a host cell.

815. **penton**

Unit of a viral capsomere that can cause hemagglutination.

816. **plaque**

Clear area in bacterial lawn resulting from lysis by bacteriophages; film of polysaccharides and bacteria on teeth.

817. **plus (positive) strand**

A viral nucleic acid having a base sequence equivalent to the viral mRNA.

818. **prion (virino)**

An infectious particle containing protein but no nucleic acid.

819. **recognition factor**

Components of viral tail fibers that bind to receptors on bacterial cells.

820. **replicative form (RF)**

Double stranded nucleic acid formed from a single stranded viral genome.

821. **retrovirus**

A RNA virus that uses its reverse transcriptase to make DNA.

822. **reverse transcriptase**

An enzyme that makes DNA from an RNA template.

823. **rise period (burst)**

Period during which the number of phages released per hour increases to a constant number.

824. **segmented genome**

A viral genome with several parts, each coding for a different polypeptide.

825. **serotype**

Strain of an organism that is serologically different from another strain.

826.   syncytium

Multinucleate mass.

827.   teratogenesis

The induction of defects in a developing embryo.

828.   tissue culture

A laboratory culture of a single tissue, which can be used cultivate and study an intracellular parasite.

829.   transcriptase

An enzyme that catalyzes transcription.

830.   tumor

Abnormal new cell growth.

831.   virion

A complete virus, usually in its extracellular phase.

832.   viroid

An infectious agent affecting some plants and consisting of single-stranded RNA.

833.   virology

The study of viruses.

834.   viropexis

A kind of phagocytosis in which naked viruses are taken into animal cells.

Eukaryotic Microbes / Parasites

835. **algin (alginic acid)**

A mucopolysaccharide found in cell walls of brown algae.

836. **algology**

The study of algae.

837. **amoeboid**

Pertaining to movement involving cytoplasmic flow.

838. **anisogametes**

Gametes of unequal size or shape.

839. **antheridium**

Male reproductive structure of sac fungus.

840. **antibiosis**

Natural production of an antimicrobial agent; against life.

841. **arthrospore**

A spore formed by fragmentation of hyphae.

842. **ascocarp**

A structure in ascomycetes containing ascospores.

843. **ascogenous hypha**

A hypha that gives rise to one or more asci.

844. **ascogonium**

Female reproductive structure in a sac fungus.

845.   ascospore

Sexual spores produced in an ascus of a sac fungus.

846.   autogamy

Union of haploid gametes from same parent organism.

847.   axopodium

Slender pseudopod associated with an axial filament.

848.   basidiocarp

Fruiting body of a basidiomycete that holds basidia.

849.   basidiospore

A spore produced by karyogamy and meiosis and found on the outside of a basidium.

850.   basidium

A club-shaped structure found in basidiomycetes.

851.   benthic

Pertaining to the bottom of the ocean or other body of water.

852.   biological vector

One living organism that carries another between hosts.

853.   cercaria

Final free-swimming larval stage of a trematode.

854.   chitin

A polysaccharide in fungal cell walls and insect exoskeletons.

855.   chlamydospore

An asexually produced spore made by some fungi.

856. **coelom**

Membrane lined body cavity of higher animals.

857. **colonial**

Pertaining to a colony of organisms.

858. **commensal**

An organism that can live on or in another organism without causing harm.

859. **conidiospore**

An asexual thin-walled spore on hyphae.

860. **conjugant**

One of complementary mating types that participate in conjugation, a kind of protozoan sexual reproduction.

861. **contractile vacoule**

A fluid-filled vacuole in some protists that moves fluid out of the organism in osmoregulation and excretion.

862. **cyst**

Spherical, thick-walled structure resembling an endospore made by some bacteria.

863. **cysticercus**

A sac with a tapeworm scolex invaginated into it.

864. **cytostome**

Permanent structure for food ingestion in a ciliate.

865. **definitive host**

Organism in which a parasite is present in its sexually mature form.

866. **diatom**

Plant-like protist with a glassy outer shell.

867. dikaryotic

Having two nuclei.

868. dimorphism

Having two different structures, usually displayed when a habitat changes.

869. dioecious

Having male and female gonads in separate individuals.

870. ectoparasite

A parasite that lives on its host's surface.

871. encystation

Cyst formation.

872. endoparasite

A parasite that lives inside its host.

873. euglenoid

A plantlike protist with one flagellum and a pigmented eyespot.

874. eutrophication

Elevation of nutrient levels due to detergents, fertilizers, and similar substances.

875. excystation

The escape of an organism from a cyst.

876. facultative parasite

A parasite that can live free of its host.

877. filamentous

Having a threadlike structure.

878. **food vacuole**

A storage reservoir for food in protists and some simple animals.

879. **frustule**

In diatoms, a cell wall containing silicon.

880. **fucoxanthin**

A brownish pigment in brown algae.

881. **gametangium**

A structure in which gametes are formed or found.

882. **gametocyte**

A male or female cell in protozoa.

883. **helminth**

A worm.

884. **hermaphroditic**

Having both male and female organs in the same body.

885. **holozoic nutrition**

The obtaining of nutrients by phagocytosis and subsequent formation of a food vacuole (phagosome).

886. **host**

Organism that harbors a parasite or infectious agent.

887. **host specificity**

Different types of organisms in which a parasite can mature.

888. **hydatid cyst**

a large cyst containing many tapeworm scolexes.

889. **hyperparasitism**

A parasite iteself having parasites.

890. **hypha**

Long threadlike structure.

891. **hypotheca**

Smaller part of a frustule of a diatom.

892. **intermediate host**

A host required for some developmental stage of a parasite.

893. **isogamete**

One of identical gametes.

894. **karyogamy**

Fusion of nuclei to produce a diploid cell.

895. **kelp**

Common name for a group of large brown algae.

896. **laminarin**

A glucose polymer stored by brown algae.

897. **lobopodium**

A lobelike pseudopodium.

898. **macrogamete**

Larger of two gametes, usually the female gamete.

899. **macronucleus**

Larger of two nuclei in ciliated protozoa.

900. **mechanical vector**

A vector in which a parasite is carried without completing any part of its life cycle.

901. **merozoite**

A trophozoite of the malaria parasite that invades erythrocytes.

902. **metacercaria**

A developmental stage of a fluke prior to transfer to a final host.

903. **microfilaria**

First juvenile stage of a filarial nematode, usually seen in blood and fluids of definitive host.

904. **micronucleus**

Smaller of two nuclei found in ciliated protozoa, which is involved in genetic recombination.

905. **miracidium**

A ciliated larval stage of a trematode, usually found in an intermediate host.

906. **monoecious**

Having male and female gonads in the same organism.

907. **mycologist**

A person who studies fungi.

908. **mycology**

The study of fungi.

909. **mycotoxicology**

The study of the effects of fungal toxins.

910. **obligate parasite**

An organism that must spend all or most of its life in or on a host.

911.   oocyst

Cyst formed around a zygote as in malaria parasites and other protozoa.

912.   oogonium

Female cell that divides by mitosis to produce a primary oocyte.

913.   osmotropic

Absorbing organic molecules in a soluble form.

914.   parasite

Organism that lives in or on and at the expense of a host.

915.   parasitology

The study of parasites.

916.   pathogen

Organism capable of causing disease.

917.   pellicle

Relatively rigid layer beneath the cell membrane in many protozoa and algae.

918.   permanent parasite

A parasite that once reaching a host remains in or on it.

919.   phycology

The study of algae.

920.   phytoplankton

A community of small floating plant-like organisms.

921.   plankton

A community of small floating organisms.

922. plasmodium

A multinucleate mass of protoplasm surrounded by a membrane characteristic of certain slime molds.

923. plasmogamy

Sexual reproduction in fungi in which haploid gametes unite and cytoplasm mixes.

924. proglottid

A tapeworm segment containing male and female sex organs.

925. pseudoplasmodium

A footlike amoeboid structure containing many mxyameba of a cellular slime mold.

926. redia

An asexually produced trematode larva.

927. reservoir host

Nonhuman organism that can harbor a pathogen and transmit it to a human.

928. saprophytic

Growing on decomposing matter.

929. saprozoic nutrition

Obtaining of nutrients in dissolved form.

930. schizogony

Multiple fission in which a single cell release many cells.

931. scolex

Head of a tapeworm with attachment structures.

932. septum

A crosswall separating two fungal cells.

933.   slime mold

A fungus-like protist with both plant and animal characteristics.

934.   sporangiopore

A fungal hypha bearing a sporangium.

935.   sporangiospore

A spore inside a sporangium.

936.   sporangium

A sac-like structure containing spores in fungi.

937.   spore coat

Protein-like layer around the cortex of an endospore.

938.   sporocyst

Larval fluke usually found in an intermediate snail or mollusk host.

939.   sporozoite

Motile infective stage of certain parasites, as in malarial parasites released from ooctes in mosquitoes.

940.   stigma

Eyespot of a euglenoid.

941.   stonewort

A group of algae with complex growth patterns found in fresh and brackish waters.

942.   swarm cell

Flagellated cell, especially a myxomycete.

943.   syngamy

Fusion of gametes (usually an egg and a sperm).

944. **temporary parasite**

A parasite that stays on its host only long enough to feed.

945. **thallus**

A plantlike body lacking leaf, stem, and root seen in some algae, fungi, and lichens.

946. **theca**

A secreted, tightly attached outer layer of a dinoflagellate.

947. **trichocyst**

Barbed tentacle-like device on ciliates for attachment or capturing prey.

948. **trophozoite**

Motile, feeding stage of a protozoan, such as malarial parasite.

949. **vector**

Organism that carries an infectious agent from one host to another.

950. **Woronin body**

Fungal organelle which blocks a septal pore and prevents matter from a damaged cell from entering a healthy one.

951. **zooplankton**

A community of small floating animals.

952. **zygospore**

A thick walled sexual spore of zygomycete fungi.

Sterilization / Disinfection

953.   aerosol

A cloud of liquid droplets in air.

954.   algicide

Agent that kills algae.

955.   antimicrobial agent

A chemotherapeutic agent used to treat microbial infections.

956.   antiseptic

Against sepsis; an agent that destroys or inhibits growth of microorganisms on tissues.

957.   autoclave

Instrument used to sterilize materials with moist heat under pressure.

958.   bactericide

An agent capable of killing bacteria.

959.   bacteriostatic

Inibiting bacterial growth and reproduction.

960.   cavitation

Formation of a cavity.

961.   D value

Decimal reduction time or time to kill 90 percent of organisms or spores at a particular temperature.

962.   decimal reduction time

Time to kill 90 percent of organisms in a population at a particular temperature.

963. **detergent**

Organic molecule that acts as a wetting agent, and sometimes as a cleansing agent.

964. **disinfectant**

A chemical agent that destroys microorganisms used on inanimate objects.

965. **disinfection**

Reducing numbers of pathogenic organisms so they pose no disease threat.

966. **fungicide**

Agent that kills fungi.

967. **germicide**

Agent that kills many microorganisms but not necessarily spores.

968. **lyophilization**

Freeze-drying to preserve microorganisms.

969. **pasteurization**

The heating of milk or other liquids to destroy pathogens or organisms that might cause spoilage.

970. **phenol coefficient test**

A test for measuring effectiveness of disinfectants by comparing them to the action of phenol on test bacteria.

971. **quaternary ammonium compound**

A cationic detergent with four units attached to nitrogen.

972. **rad**

A unit of absorbed radiation equal to 100 ergs of energy.

973. **sanitization**

Decreasing microbial populations on inanimate objects to levels judged safe by public health standards.

974.   sonication

The use of sound waves to break cells apart.

975.   sterility

Absence of living organisms.

976.   sterilization

The killing or removing of all living organisms in or on a substance.

977.   thermal death point

Lowest temperature that can kill all organisms in a microbial suspension in ten minutes.

978.   thermal death time

Shortest time needed to kill all organisms in a microbial suspension at specified temperature and conditions.

979.   tincture

A solution in which alcohol is the solvent.

980.   tyndallization

Fractional steam sterilization.

981.   use-dilution test

A procedure for evaluating chemotherapeutic agents using pharmaceutical preparations and certain test bacteria.

982.   viricide

An agent that inactivates viruses so they cannot replicate in cells.

983.   wetting agent

A detergent that causes oil and water to mix, making it possible for other agents to penetrate fatty substances.

Antimicrobial Therapy

984. **aminoglycoside antibiotic**

An antibiotic with a cyclohexane ring that binds to small ribosomal units and inhibits protein synthesis.

985. **antibiotic**

A substance made by a microorganism that destroys or inhibits growth of other microorganisms.

986. **antibiotic resistance**

Development of ability to resist action of an antibiotic.

987. **antifungal agent**

Agent that kills or inhibits growth of a fungus.

988. **antimetabolite**

A substance that interferes with normal metabolic pathways.

989. **broad spectrum antibiotic**

A chemotherapeutic agent with a wide range of activity.

990. **cephalosporin**

A chemotherapeutic agent with a beta-lactam structure derived from the fungus Cephalosporium.

991. **chemotherapeutic agent**

A chemical substance used to treat disease.

992. **chemotherapeutic index**

A measure of tolerable dose of a drug divided by the minimum dose to cure a disease (both per unit body weight).

993. **chemotherapy**

Use of chemical agents to treat disease.

994. **cross resistance (to drugs)**

Ability to avoid damage by two or more drugs because of the same mechanism.

995. **dilution method**

Procedure for testing antibiotic sensitivity of organisms.

996. **disk diffusion (Kirby-Bauer) method**

Method of determining an organisms sensitivity to antimicrobial agents by placing antibiotic disks on culture.

997. **extrachromosomal drug resistance**

Resistance to an antibiotic due to plasmids not incorporated into the bacterial chromosome.

998. **growth factor**

An organic substance that stimulates growth.

999. **interferon**

One of a few glycoproteins from virus-infected cells that stimulate adjacent cells to produce antiviral protein.

1000. **macrolide antibiotic**

An antibiotic chemically characterized by the presence of a macrolide (modified lactone) ring.

1001. **minimum bactericidal concentration**

Lowest concentration of an antimicrobial agent that kills organisms as shown by absence of growth in cultures.

1002. **minimum inhibitory concentration**

Lowest concentration of an antimicrobial agent that prevents growth in an antibiotic sensitivity test.

1003. **minimum lethal concentration**

Lowest concentration of an antimicrobial agent that will kill a particular microorganism.

1004. **molecular mimicry**

Imitation of a normal molecule by an antimetabolite.

1005. narrow spectrum antibiotic

A chemotherapeutic agent that attacks only a few species of microorganisms.

1006. parenteral route

A route of drug administration other than by mouth.

1007. selective toxicity

Ability of a chemotherapeutic agent to kill or inhibit microorganisms without damaging host cells.

1008. semisynthetic drug

An antibiotic made partly by a microorganism and partly by laboratory procedures.

1009. serum killing power

Test for effectiveness of an antimicrobial agent by adding bacteria to serum of patient receiving antibiotic.

1010. superinfection

A new infection caused by a bacterium or fungus resistant to a drug being used to treat the patient.

1011. synergistic effect

Positive or negative effect of two agents greater than the individual effects of either agent.

1012. synthetic drug

A chemotherapeutic agent made in the laboratory.

1013. therapeutic dosage level

Amount of a chemotherapeutic agent that will eliminate a pathogen over a period of time.

1014. zone of inhibition

Clear area on agar showing where an agent on a disk has inhibited growth of a microorganism.

Host, Microbe and Disease

1015. abortive infection

Viral infection in which viruses enter cells but cannot make progeny.

1016. acute-phase reactant

Qualitative and quantitative changes in nonspecific defenses in host blood early in an infection.

1017. adherence

Attachment of a microorganism to a host cell surface.

1018. adhesin

A substance that aids a microorganism in attaching to a host cell.

1019. antitoxin

An antibody to a particular toxin.

1020. chronic disease

A disease that develops slowly and persists for a long, indeterminate period.

1021. coagulation

Blood clot formation.

1022. colonization

Formation of colonies.

1023. comedo

A dried sebum plug in an oil gland on the skin.

1024. commensalism

Symbiotic relationship in which one organism benefits and the other one neither benefits nor is harmed.

1025. communicable disease

Infectious disease passed from one host to another.

1026. contagious disease

A disease that can spread from one host to another.

1027. contamination

Introduction of unwanted microorganisms.

1028. convalescence

Recovery phase of an illness.

1029. critical stage (acme of infection)

Period of most intense symptoms.

1030. crustose

In lichens, being compact and pressed to substrate.

1031. cytocidal

Killing a host cell.

1032. decline phase (of infection)

Period during which defenses overcome a pathogen and symptoms subside.

1033. ectosymbiosis

A symbiotic relationship in which one organism remains outside the other.

1034. endogenous disease

Disease caused by a species of an organism's own flora.

1035. endogenous pyrogen

The lymphokine, interleukin-1, which induces fever.

1036.   endosymbiosis

The presence of one organism inside another.

1037.   endotoxin

A heat stable lipopolysaccharide in the outer membrane of a gram-negative bacterial cell wall, toxic on cell lysis.

1038.   enterotoxin

A toxin that specifically affects intestinal mucosal cells.

1039.   eructation

Belching forth.

1040.   exogenous disease

Disease caused by organisms entering the body from the environment.

1041.   exotoxin

A toxin secreted by microorganisms into their environment, including the tissues of a host.

1042.   fever

Abnormally high body temperature.

1043.   fibrinolysin

Enzyme from streptococci tha digests fibrin and thus blood clots; streptokinase.

1044.   final host

The host in which a parasites gains sexual maturity or reproduces.

1045.   focal infection

An infection confined to a particular area from which organisms can spread.

1046.   foliose

Leaflike appearance, as of lichens.

1047. fomite

Nonliving object capable of transmitting disease.

1048. fruticose

A shrubby shape, as in lichens.

1049. gnotobiotic

Living germfree or in an environment with only a few known microorganisms.

1050. granulocytopenia

A shortage of blood granulocytes.

1051. hemolysin

An agent that causes erythrocyte lysis.

1052. hyaluronidase

A bacterial enzyme that digests host tissue hyaluronic acid and causes cells to separate making invasion easier.

1053. iatrogenic disease

A disorder caused by medical treatment.

1054. idiopathic disease

A disorder with no known cause.

1055. incubation period

The time between infection and the appearance of disease symptoms.

1056. infection

Multiplication of an organism (usually a microorganism) within the body.

1057. infectious disease

A disease caused by an infectious agent.

1058. **infestation**

Presence of arthropod, helminth, or protozoan parasites on or in a living host.

1059. **inflammation**

Localized response to tissue injury usually involving pain, redness, heat, and swelling.

1060. **intoxication**

Affected by the presence of a toxin.

1061. **invasive phase**

Period during which signs and symptoms of a disease appear.

1062. **invasiveness**

Ability of a microorganism to enter, grow, and reproduce in a host.

1063. **latent disease**

A disease with a long period between infection and the first appearance of signs and symptoms.

1064. **lethal dose 50**

Dose of toxin or number of microorganisms that will kill 50 percent of an experimental group in a given time.

1065. **leukocidin**

Toxin made by staphylococci that kills phagocytes.

1066. **leukostatin**

Toxin from streptococci and staphylococci that keeps leukocytes from engulfing these organisms.

1067. **lichen**

A symbiotic association of an alga and a fungus.

1068. **local infection**

Infection found in a specific part of the body.

1069. microbial antagonism

One species producing a substance that interferes with the growth of another species.

1070. microbiota

Microorganisms associated with a particular structure.

1071. mixed infection

An infection to which several species of causative agents contribute.

1072. mononuclear phagocyte system

The body collection of phagocytic cells.

1073. mucociliary blanket (escalator)

A layer of mucus and cilia lining parts of the respiratory system and trapping microorganisms.

1074. mutualism

A symbiotic relationship that benefits both of the involved organisms.

1075. mycobiont

The fungal part of a lichen.

1076. neurotoxin

A toxin that acts on the nervous system.

1077. noncommunicable infectious disease

A disease caused by an infectious agent but not spread from one host to another.

1078. noninfectious disease

Disease caused by something other than an infectious agent.

1079. normal flora

Microorganisms commonly found in or on another organism.

1080. opportunist

A usually nonpathogenic organism that can under certain circumstances cause disease.

1081. parasitism

A symbiotic relationship in which one organism damages another but cannot live without it.

1082. pathogenicity

Ability to cause disease.

1083. peristalsis

Rhythmic contractions that push substances along the digestive tract.

1084. phagolysosome

A particle formed from the fusion of a phagosome and lysosomes.

1085. phagosome

A particle formed by phagocytosis.

1086. phycobiont

The algal component of a lichen.

1087. primary infection

An initial infection in a previously healthy individual.

1088. prodromal phase

A short period of nonspecific symptoms at the onset of a disease.

1089. pyrogen

Agent that causes fever.

1090. resident flora

Microorganisms always present in or on an organism.

1091. **reticuloendothelial system**

The body's aggregate of phagocytic cells.

1092. **sapremia**

A condition in which saprophytes release metabolic products into blood.

1093. **secondary infection**

Infection that follows a first infection, especially in debilitated patients.

1094. **septicemia**

Presence of rapidly multiplying pathogens in blood; blood poisoning.

1095. **sign**

A characteristic of a disease observable by examining a patient.

1096. **streptokinase**

Streptococcal enzyme that digests fibrin thereby dissolving blood clots.

1097. **streptolysin-O**

Hemolysin that causes beta hemolysis of blood cells on agar plates under anaerobic conditions.

1098. **streptolysin-S**

Hemolysin that can kill phagocytes and cause beta hemolysis of blood cells on agar plates under aerobic conditions.

1099. **subacute disease**

A disease with effects intermediate between chronic and acute.

1100. **subclinical infection**

Infection that fails to produce symptoms; inapparent infection.

1101. **symbiont**

An organism living in a symbiotic relationship.

1102. **symbiosis**

A close association between dissimilar organisms.

1103. **symptom**

An effect of a disease felt by patient but not directly observable, such as nausea and pain.

1104. **syndrome**

A set of signs and symptoms seen together in a particular disease.

1105. **systemic infection**

Infection that affects the whole body.

1106. **toxemia**

A condition produced by toxins being present in a host's blood.

1107. **toxigenicity**

Ability of an organisms to produce a toxin.

1108. **toxin**

A microbial product that can injure another organism at low concentration; a poisonous substance.

1109. **toxoid**

A bacterial exotoxin modified so that it retains antigenic properties but is no longer poisonous.

1110. **transfer host**

Host that can carry a parasite to its final host but that is not needed to complete the life cycle.

1111. **transient flora**

Microorganisms present in or on an organism under certain conditions and times.

1112. **viremia**

Presence, but not replication, of viruses in the blood.

1113. virulence

The degree of pathogenicity or intensity of a disease produced by a pathogen.

Epidemiology / Nosocomial Infections

1114. active carrier

A host that releases disease organisms long after recovery.

1115. airborne transmission

Transmission of an infectious organism through air for more than a meter from source to host.

1116. autogenous infection

Infection from a patient's own flora.

1117. casual carrier

An individual who harbors an infectious organism for a short time period.

1118. cell-mediated immunity

Specific immunity carried out mainly by T lymphocytes.

1119. Centers for Disease Control

Branch of the U. S. Public Health Service concerned with prevention and control of disease.

1120. chronic carrier

An individual who harbors a pathogen for a long time period.

1121. common source epidemic

Epidemic arising from contact with a contaminated substance.

1122. direct contact transmission

Mode of disease transmission involving body contact.

1123. endemic disease

A disease nearly always present and at a relatively low but steady frequency in a population.

1124.   enzootic

Moderate prevalence of a disease in an animal population.

1125.   epidemic

A disease that displays a sudden increase in occurrence over the normally expected number of cases in a population.

1126.   epidemiologist

A person who practices epidemiology.

1127.   epidemiology

The study of factors in the frequency and distribution of diseases and health related events in a human population.

1128.   epizootic

A sudden outbreak of a disease in an animal population.

1129.   etiology

The causes and origins of a disease.

1130.   fecal-oral transmission

The carrying of microorganisms from feces to the mouth.

1131.   harborage transmission

Transfer of an infectious organism without being changed while within the vector.

1132.   healthy carrier

An individual who harbors a pathogen without being ill.

1133.   herd immunity

Proportion of individuals in a population who are immune to a particular disease.

1134.   horizontal transmission

Disease transmission by direct contact.

1135. **hyperendemic disease**

Disease showing a gradual increase in incidence over the endemic level in a population; endemic to all age groups.

1136. **incidence**

The number of new cases (of a disease) in a specific time.

1137. **incubatory carrier**

An individual carrying an infection organism before symptoms appear.

1138. **index case**

First disease case in an epidemic in a given population.

1139. **intermittent carrier**

An individual that releases disease organisms periodically.

1140. **isolation**

Prevention of contact between an individual with a communicable disease and general population

1141. **morbidity rate**

The number of individuals contracting a disease in relation to susceptible population over a specific time period.

1142. **mortality rate**

The number of individuals dying from a disease in relation to the total number of cases of a disease.

1143. **nosocomial infection**

Infection acquired in a medical facility.

1144. **notifiable disease**

A disease a physician must report to a public health department.

1145. **outbreak**

Sudden occurrence of a disease in a population.

1146. **pandemic**

Increase in occurrence of a disease over a very wide area; worldwide epidemic.

1147. **panzootic**

Wide dissemination of a disease among an animal population.

1148. **passive carrier**

An individual who releases infectious organisms without having been affected by them.

1149. **placebo**

An inactive substance given to some recipients to test the effect of certain medications or treatments.

1150. **portal of entry**

A site at which a microorganism can enter a host.

1151. **portal of exit**

A site at which a microorganism can leave a host.

1152. **prevalence**

The number of individuals infected with a particular organism at any time.

1153. **propagated epidemic**

An epidemic arising from person-to-person transmission.

1154. **quarantine**

Separation of humans or animals with an infectious disease from healthy members of the population.

1155. **reservoir of infection**

Site where microorganisms can maintain their ability to infect.

1156. **sporadic disease**

A disease limited to a small population with little risk of its spreading to a large population.

1157.   vector control

Control of the spread of a disease by controlling its vector.

1158.   vector-borne transmission

Disease transmission between hosts by a vector.

1159.   vertical transmission

Passage of pathogens from parent to offspring, as in egg or sperm, across placenta, or in birth canal.

1160.   zoonosis

A disease that can be passed from animal to human.

Immunology

1161. **ABO blood group**

Major human blood types based on cell membrane A and B antigens.

1162. **acquired immune tolerance**

Ability to produce antibodies against nonself antigens but not to produce them against self antigens.

1163. **acquired immunity**

Specific immunity obtained by a means other than heredity.

1164. **active immunity**

Immunity obtained by the host's body producing antibodies or other defenses against a disease agent.

1165. **adjuvant**

A substance added to an antigen that increases its immunogenicity.

1166. **agglutinate**

Aggregate by agglutination.

1167. **agglutination reaction**

Formation of an immune complex of antigens and antibodies with attached cells or other particles.

1168. **agglutinin**

Antibody responsible for agglutination.

1169. **agglutinogen**

Antigen that initiates agglutination.

1170. **AIDS-related complex**

A set of symptoms associated with HIV infection.

1171. allergen

Normally innocuous substance that can bring about an adverse inmmunologic response.

1172. allergy

Immune system disorder involving an inappropriate response, usually to an antigen that is normally ignored.

1173. allograft

Graft between genetically different members of same species.

1174. allotype

Allelic variant of an antigenic determinant.

1175. anamnestic response

A prompt immunological response because of immunologic memory of previous exposure to an agent.

1176. anaphylaxis

Immediate hypersensitivity (Type I) of previously sensitized individual.

1177. antibody

A protein that is capable of binding to a specific antigen.

1178. antibody-mediated immunity

Humoral immunity.

1179. antigen

Any foreign substance against which the body produces an immune response.

1180. antigen-binding fragment

The portion of an antibody molecule that binds with a specific antigenic determinant.

1181. antigenic determinant site

Epitope.

1182. **antigenic drift**

Antigenic variation that creates a strain unrecognizable by host immunity or a new neuraminidase or hemagglutinin.

1183. **antigenic shift**

Antigenic variation thought to be due to reassortment of viral genes.

1184. **antihistamine**

Drug that counteracts the effects of histamine.

1185. **antiserum**

Serum that contains induced antibodies.

1186. **antiviral protein**

A protein released by a virally infected cell that prevents viral replication in adjacent cells.

1187. **Arthus reaction**

A local hypersensitivity reaction involving formation of an immune complex.

1188. **artificially acquired active immunity**

Specific immunity obtained from the body's response to antigens or toxins in vaccines.

1189. **artificially acquired passive immunity**

Specific immunity obtained from receiving ready made antibodies or antitoxins.

1190. **atopy**

Allergic reactions occuring first at the site of the allergen's entry into the body.

1191. **autoallergy**

Autoimmune disease.

1192. **autograft**

Graft of tissue from one part of the body to another.

1193. autoimmune disease

Immune response to one's own tissue.

1194. B lymphocyte (B cell)

Lymphocyte derived from bone marrow and involved in making antibodies.

1195. blocking antibody

IgGs elicited in allergy patients by gradually increasing allergen doses to prevent binding of allergen with IgEs.

1196. blood type

A characteristic of blood based on the nature of antigens on erythrocyte membranes.

1197. bradykinin

A kinin that may elicit pain following tissue injury.

1198. chromogen

Colorless substrate that after being acted on by an enzyme forms a colored product.

1199. chronic inflammation

A condition characterized by a persistent contest between an inflammatory agent and phagocytes and other defenses.

1200. classical complement pathway

Pathway for activating complement that requires antibodies.

1201. clonal selection theory

A theory that clones of lymphocytes arise from single cells and reproduce when an antigen binds to their receptors.

1202. colostrum

Mammary gland secretion that forms prior to milk production.

1203. complement fixation

Activation of complement and its subsequent removal from blood.

1204. compromised host

Host with lowered resistance to disease.

1205. constant region

Portion of an immunoglobin having nearly the same structure for immunoglobulins of the same type.

1206. convalescent serum

Serum having high concentration of antibodies depending on the disease from which donor was recovering.

1207. cytokine

Any nonantibody protein released in response to an antigen.

1208. cytotoxic T cell

Cell that can recognize and destroy virus-infected cells.

1209. degranulation

Release of histamine and other mediators of allergy on second exposure; formation of phagolysosomes.

1210. delayed-type hypersensitivity

Cell-mediated hypersensitivity; reaction delayed hours to days after exposure.

1211. desensitization

Rendering an individual with an allergy nonreactive to the allergen.

1212. diapedesis

Passage of leukocytes between cells of capillary walls in an inflammatory reaction.

1213. double diffusion (Ouchterlony) assay

Immunodiffusion reaction in which antigen-antibody complexes become visible.

1214. effector T cell

A kind of lymphocyte that directly attacks specific target cells.

1215. **enzyme-linked immunosorbent assay (ELISA)**

Test in which an antiantibody is attached to enzyme that causes a color change in its substrate.

1216. **epitope**

Area of an antigen that stimulates and combines with a specific antibody; antigenic determinant site.

1217. **exogenous pyrogen**

Toxin from an infectious agent that produces fever in the host's body.

1218. **fibroblast**

Connective tissue cell that produces fibrin and other fibrous proteins.

1219. **gamma globulin**

A pooled sample of the antibody-containing fraction of serum.

1220. **generalized anaphylaxis**

Hypersensitivity (Type I) appearing in a life-threatening manner.

1221. **genetic immunity**

Specific disease resistance because of genetic factors.

1222. **graft-vs-host disease**

Disease in which host antigens elicit antibody production by graft cells thereby destroying tissue.

1223. **granuloma**

In inflammation, an aggregate of dead tissue, phagocytes, and lymphocytes.

1224. **granulomatous inflammation**

Generalized reaction with monocytes, lymphocytes, plasma cells, and histiocytes present.

1225. **gut-associated lymphatic tissue (GALT)**

Collective name for lymphatic tissue in the digestive tract.

1226. **hapten**

A molecule that becomes immunogenic when attached to another molecule.

1227. **helper T cell**

A cell that assist in presenting antigens to B lymphocytes.

1228. **hemagglutination**

Agglutination of erythrocytes by antibodies.

1229. **hemagglutinin**

An antibody that causes hemagglutination.

1230. **heterogeneity**

Having many different kinds, as in the immune system producing many different kinds of antibodies.

1231. **heterologous antigen (heterophile)**

An antigen that cross-reacts with antibodies produced by a different antigen, usually because of similar epitopes.

1232. **histamine**

An amine released during allergic reactions.

1233. **histocompatibility antigen**

A cell surface antigen recognized by the immune system and thus important in transplant rejection.

1234. **host-vs-graft disease**

Disease in which graft antigens elicit antibody production by host cells, thereby destroying graft tissue.

1235. **human leukocyte antigen (HLA)**

Antigens on lymphocytes used to determine compatibility of donor and recipient tissues.

1236. **humoral immunity**

Specific immunity produced by antibodies.

1237. hyperimmune serum

A gamma globulin containing large numbers of a particular kind of antibody.

1238. hypersensitivity

Inappropriate response of the immune system such as to an antigen it normally ignores.

1239. IgA

Immunoglobulin present in body secretions.

1240. IgD

Surface immunoglobulin on B lymphocytes that may serve as an antigen receptor in stimulating antibody synthesis.

1241. IgE

Immunoglobulin that binds to mast cells and basophils in allergic reactions.

1242. IgG

Immunoglobulin present in blood; antibody.

1243. IgM

First immunoglobulin formed in an infection, released as a five-unit molecule,

1244. immnoglobulin

A Y-shaped protein containg four polypeptides, which can combine with a particular antigen; antibody.

1245. immune adherence

Opsonization.

1246. immune complex

Product of an antigen-antibody reaction, which can also contain complement.

1247. immune response

The body's response to the presence of an antigen.

1248. immune surveillance

A proposed process by which immune system killer cells identify and destroy tumor cells and virus-infected cells.

1249. immunity

Organism's ability to recognize infectious agents and defend against them.

1250. immunodeficiency

Inadequate response by the immune system to antigens because of acquired or inborn defects in B or T cells.

1251. immunodiffusion

Procedure for serological tests in solid media.

1252. immunoelectrophoresis

Procedure for separating antigens by electrophoresis before reacting them with antibodies.

1253. immunofluorescence

Procedure for combining antibodies with fluorescent dyes for ease of identification.

1254. immunogen

Antigen.

1255. immunology

The study of specific defense mechanisms.

1256. immunoprecipitation

Reacting antigens and antibodies to form large particles that precipitate out of solution.

1257. immunosuppression

Decreasing the response of the immune system with radiation or cytotoxic drugs, as to prevent transplant rejection.

1258. immunotoxin

A monoclonal antibody attached to a specific toxin for use in destroying cancer and other target cells.

1259. induration

Red, hard elevation on skin, as from sensitivity to tuberculin.

1260. innate immunity

Genetically determined immunity.

1261. interleukin

One of a few proteins from macrophages and T cells that regulate lymphocytes in humoral and cellular immunity.

1262. isograft

Graft between genetically identical individuals (identical twins).

1263. J chain

Polypeptide that links units of IgA or IgM immunoglobulins.

1264. jaundice

A yellowing of tissues from bilirubin deposition.

1265. killer (K) cell

Killer T cell growing in vitro.

1266. Lancefield system

System for serologically distinguishing groups of streptococci.

1267. lymphokine

A protein secreted by an activated lymphocyte.

1268. macrophage

Large phagoctic cell that also activates lymphocytes.

1269. major histocompatibility complex (MHC)

A set of genetically determined cell surface antigens specific to each individual.

1270. **membrane attack complex (MAC)**

Complement components that create pores in plasma membranes and cause cell lysis.

1271. **memory cell**

A lymphocyte sensitized to a particular antigen that remains to detect the antigen if it again enters the body.

1272. **migration inhibition factor (MIF)**

A lymphokine from T cells that inhibits macrophage movement and fosters macrophage accumulation in the area.

1273. **mitogen**

An agent that induces mitosis.

1274. **myeloma cell**

A tumor cell similar to bone marrow cells; a plasma cell easily cultured to make large quantities of antibodies.

1275. **natural killer (NK) cell**

Lymphocyte that destroy malignant and transplanted cells.

1276. **naturally acquired active immunity**

Specific immunity due to an organism's response to an infectious agent.

1277. **naturally acquired passive immunity**

Specific immunity due to an organism receiving antibodies made by another organism.

1278. **nonspecific defense**

A defense against microorganisms that operates regardless of invading agent.

1279. **nonspecific immunity (resistance)**

Immunity produced by nonspecific defenses.

1280. **null cell**

A lymphoid cell lacking properties of either T or B cells.

1281. oncology

The study of cancer and tumors.

1282. opsonin

Antibody bound to a microorganism and fostering its phagocytosis.

1283. opsonization

The rendering of microorganisms more attractive to phagocytes by coating with opsonins and complement.

1284. passive immunity

Specific immunity acquired be receiving antibodies from another organism.

1285. perforin

A lethal protein made by cytotoxic and natural killer cells that makes holes in target cell membranes.

1286. Peyer's patch

An aggregation of lymphatic tissue in small intestine.

1287. plasma cell

A mature, sensitized B lymphocyte that makes and secretes antibodies.

1288. preciptin

An antibody that causes a precipitation reaction.

1289. preciptin reaction

Formation of an insoluble precipitate by an antigen antibody complex.

1290. primary immunodeficiency

Deficient immune response due to a genetic or developmental defect.

1291. properdin pathway

An alternative pathway to classic complement fixation.

1292. **prostaglandin**

A fatty acid derivative that acts as a cellular regulator.

1293. **Quellung reaction**

Swelling of a capsule in presence of antibodies against capsular antigens.

1294. **radial immunodiffusion (RID)**

Serological test to measure concentrations of antigen and antibody.

1295. **radioallergosorbent test (RAST)**

Radioimmunoassay used to measure IgE directed toward an allergen.

1296. **radioimmunoassay (RIA)**

Use of radioisotope-labeled antigen/antibody to compete with its unlabeled counterpart to determine concentrations.

1297. **radioimmunosorbent test (RIST)**

Radioimmunoassay used to measure IgE remaining in body fluids.

1298. **reagin**

Antibody (IgE) that mediates immediate hypersensitivity reactions.

1299. **regulator T cell**

Lymphocyte that controls development of effector T cells.

1300. **resistant**

Able to defend against.

1301. **Rh system**

A set of antigens on erythrocytes responsible for hemolytic disease of the newborn.

1302. **scab**

A covering over a skin injury made of dried fibrin and trapped cells.

1303. Schick test

A serological test involving neutralization that detects antitoxin or antibodies to diphtheria toxin.

1304. secondary immunodeficiency

Deficient immune response due to damage to lymphocytes after normal development.

1305. secretory IgA

Immunoglobulin found in body fluids.

1306. self-vs-nonself

Distinguishing between molecules that are part of the body and those foreign to it.

1307. serology

Branch of immunology concerned with laboratory work with antigens and antibodies.

1308. seroreversion

A change in serum such that antibodies formerly present are no longer found.

1309. serotyping

A serological procedure to differentiate among strains of an organisms.

1310. slow-reacting substance of anaphylaxis

A mediator of allergy consisting of leukotrienes that causes slow, long-lasting airway constriction in animals.

1311. specific defense

Defense against a particular pathogen.

1312. specific immune response

A particular response of the immune system to a certain pathogen.

1313. suppressor T cell

A lymphocyte that suppresses an immunologic reaction.

1314. **susceptible**

Vulnerable to infection.

1315. **T lymphocyte (T cell)**

A lymphocyte activated in the thymus and involved in cell-mediated immunity.

1316. **T-dependent antigen**

An antigen that requires a helper T cell to present it to a B lymphocyte.

1317. **T-independent antigen**

An antigen that does not require a helper T cell to present it to a B lymphocyte.

1318. **thymocyte**

A lymphocyte in the thymus.

1319. **titer**

Quantity of substance required for a given effect; reciprocal of highest antiserum dilution with a positive result.

1320. **tolerance**

A state in which an agent, such as an antigen, no longer elicits an effect.

1321. **toxin neutralization**

Inactivation of a toxin by an antibody called an antitoxin.

1322. **transplant rejection**

Destruction of transplant or graft tissue by the host's immune system.

1323. **transplantation**

Moving of tissue from one site to another.

1324. **type I hypersensitivity**

Immediate hypersensitivity from IgE antigens causing an anaphylactic reaction.

1325. **type II hypersensitivity**

Immediate hypersensitivity due to binding of antigens and antibodies with cytotoxic (target cell) destruction.

1326. **type III hypersensitivity**

Immediate hypersensitivity due to deposition of antigen-antibody complexes, with inflammation and tissue damage.

1327. **type IV hypersensitivity**

Delayed hypersensitivity a day or two after exposure that activates T cells, causing inflammation and tissue damage.

1328. **vaccine**

A material containing an antigen that elicits an immunologic response.

1329. **variable region**

Portion of an antibody having a specific structure and able to bind with a particular antigen.

1330. **viral hemagglutination**

Clumping together of viruses with erythrocytes.

1331. **viral neutralization**

Binding of antibodies to viruses that renders them harmless.

1332. **Widal test**

A test for agglutination of typhoid bacilli when mixed with serum containing typhoid antibodies.

1333. **xenograft**

A tissue graft from one to another animal species.

Human Anatomy & Disease

1334. agranulocyte

A leukocyte lacking cytoplasmic granules.

1335. alveolus

Saclike structure.

1336. arteriole

Blood vessel that carries blood from an artery to capillaries.

1337. artery

A blood vessel that receives blood from the heart or other large artery.

1338. atrium

A chamber of the heart that receives blood from veins.

1339. basement membrane

Nonliving barrier made by epithelial and dermal cells of skin.

1340. basophils

Leukocytes that become mast cells after migrating into tissues.

1341. blood-brain barrier

A barrier created in the brain by specialized capillaries that prevent many substances from entering brain cells.

1342. bronchial tree

Branching set of passageways in the respiratory system.

1343. bronchiole

Small passageways in the respiratory system.

1344. bronchus

Large passageways in the respiratory system.

1345. capillary

Small blood vessel through which substances are exchanged with cells.

1346. cardiovascular system

Heart, blood vessels, and blood.

1347. cementum

Hard material on outer surface of a tooth below the gumline.

1348. central nervous system

Brain and spinal cord.

1349. cerebrospinal fluid

Fluid secreted into cavities inside the brain and spinal cord.

1350. cerumen

Earwax.

1351. ceruminous gland

Gland that produces earwax.

1352. colon

Most of the large intestine.

1353. conjunctiva

Mucous membrane of the eyes.

1354. cornea

Transparent anterior portion of eyeball.

1355. **crown (of a tooth)**

Part of tooth above the gumline.

1356. **dermis**

Layer of skin beneath the epidermis.

1357. **digestive tract**

Tube along which food passes as it is broken down and absorbed.

1358. **ear canal**

Passageway from the outer to middle ear.

1359. **electrolyte**

A substance that ionizes in solution.

1360. **enamel**

A hard mineralized substance that forms the outer surface of the crown of a tooth.

1361. **endocardium**

The inner layer of the heart.

1362. **endometrium**

The mucous membrane lining of the uterus.

1363. **eosinophils**

Leukocytes that are released in large quantities during an allergic reaction.

1364. **epicardium**

Outer layer of the heart.

1365. **epidermis**

Outer layer of skin.

1366. epiglottis

Structure that prevents food and fluids from entering the larynx.

1367. erythrocyte

Red blood cell.

1368. external genitalia

Sex organs outside the body.

1369. feces

Solid wastes stored in and released from the large intestine.

1370. formed element

A solid structure in blood; erythrocytes, leukocytes, and platelets.

1371. ganglion

An aggregation of neuron cell bodies.

1372. genitourinary (urogenital)

Pertaining to the reproductive and urinary tracts.

1373. glomerulus

A kidney structure through which materials are filtered from the blood.

1374. hemoglobin

An iron-containing pigment that carries oxygen in blood.

1375. keratin

Water-proofing protein in epidermal cells.

1376. kidney

Organ that excretes nitrogenous wastes and adjusts concentrations of substances in blood.

1377. **Kupffer cell**

Phagocytic cells that remove foreign matter as blood passes through sinusoids.

1378. **lacrimal gland**

Tear-producing gland.

1379. **large intestine**

Portion of the digestive tract where feces form.

1380. **larynx**

Voice box.

1381. **leukocyte**

White blood cell.

1382. **lower respiratory tract**

Portion of the respiratory system characterized by small passageways and respiratory membranes.

1383. **lymph**

Tissue fluid.

1384. **lymph capillary**

Small tube that collects tissue fluid from around cells.

1385. **lymph node**

Encapsulated lymphatic tissue that helps to remove micro-organisms from lymph.

1386. **lymphatic vessel**

Large tube carrying lymph toward blood.

1387. **lymphocytes**

White blood cells found in large numbers within lymphoid tissues.

1388. **mammary gland**

Milk producing gland.

1389. **mast cell**

A connective tissue cell that releases histamine and other active substances in inflammatory and allergic reactions.

1390. **mastoid area**

Skull bone region behind ear in which infection can occur.

1391. **megakaryocyte**

Large bone marrow cell that gives rise to platelets.

1392. **meninges**

Membranes surrounding the brain and spinal cord.

1393. **microvillus**

A very small fold in the cell membrane that increases the surface of an intestinal villus.

1394. **middle ear**

Chamber in which ear ossicles are located and often the site of infections.

1395. **monocyte**

A large white blood cell with a nonlobed nucleus.

1396. **mucin**

A glycoprotein in mucus that coats bacteria and prevents them from invading the body.

1397. **myocardium**

The muscular layer of the heart.

1398. **nasal cavity**

Portion of respiratory passageways through which air first passes.

1399. nephron

The functional unit of kidneys.

1400. nerve

An aggregation of nerve fibers (mainly axons).

1401. neuron

A signal transmitting cell of the nervous system.

1402. neutrophil

A white blood cell with a lobed nucleus and granules that stain with basic dyes.

1403. nongonococcal urethritis (NGU)

Chlamydia trachomatis, Gardnerella vaginalis, Ureaplasma urealyticum.

1404. oral cavity

Region through which food enters the digestive tract.

1405. ovarian follicle

Cluster of cells containing an ovum in a mammalian ovary.

1406. ovary

Female organ that produces ova.

1407. pericardial sac

A tough, membranous sac around the heart.

1408. peridontium

Tissue surrounding and supporting the teeth.

1409. peripheral nervous system

Cranial and spinal nerves and their branches.

1410. **pharynx**

Common passageway for respiratory and digestive tracts; throat.

1411. **pinna**

External flaplike part of ear.

1412. **plasma**

Liquid portion of blood.

1413. **platelet**

A fragment of a megakaryocyte that contains blood clotting factors.

1414. **pleura**

Membranes that cover lungs and line pleural cavities.

1415. **polymorphonuclear leukocyte (PMNL)**

A leukocyte with an irregular nucleus and granular cytoplasm.

1416. **prostate gland**

Male gland that contributes secretions to semen.

1417. **sebaceous gland**

Skin gland that secretes sebum.

1418. **sebum**

An oily secretion of a sebaceous gland.

1419. **semen**

Sperm and fluid from the male reproductive system.

1420. **serous membrane**

A membrane that secretes a watery fluid, such as the pleura.

1421.   sinus

Passageway in tissue often lined with phagocytes.

1422.   sinusoid

Enlarged capillary.

1423.   small intestine

Portion of intestine where absorption occurs.

1424.   spleen

Large lymphatic organ that filters blood.

1425.   testis

Male gonads.

1426.   thymopoietin

A hormone from the thymus thought to foster development of T cells from lymphocytes.

1427.   thymosin

A hormone from the thymus gland.

1428.   thymus gland

Lymphatic gland near the sternum that converts lymphocytes to T cells.

1429.   trachea

Main respiratory passageway; windpipe.

1430.   tympanic membrane

Membrane to which sound vibrations are transmitted from air.

1431.   upper respiratory tract

Larger nonrespiratory passageways of respiratory system.

1432. **ureter**

Tube leading from kidney to bladder.

1433. **urethra**

Tube leading from bladder outside the body.

1434. **urinalysis**

Laboratory analysis of urine.

1435. **urinary bladder**

Baglike structure where urine collects until urination occurs.

1436. **uterine tube**

Passageway that carries ova from an ovary to the uterus.

1437. **uterus**

Hollow organ where a fetus develops.

1438. **vagina**

Passageway from the uterus outside the body.

1439. **vein**

A blood vessel that carries blood toward the heart.

1440. **ventricle**

A chamber of the heart that forces blood into blood vessels.

1441. **venule**

A blood vessel that carries blood from capillaries to veins.

1442. **villus**

A multicellular extension that increases the surface area of a mucous membrane, facilitating absorption.

Diseases and Organisms

1443. **acquired immune deficiency syndrome (AIDS)**

HIV virus.

1444. **actinomycosis**

Actinomyces israelii.

1445. **acute infantile gastroenteritis**

Rotavirus.

1446. **AIDS**

Human immunodeficiency virus (HIV). (Same as term number 1443.)

1447. **amoebic dysentery**

Entamoeba histolytica.

1448. **anthrax**

Bacillus anthracis.

1449. **ascariasis**

Ascaris lumbricoides.

1450. **aspergillosis**

Aspergillus sp.

1451. **atypical pneumonia**

Mycoplasma pneumoniae.

1452. **avian tuberculosis**

Mycobacterium avium.

1453. babesiosis

Babesia microti

1454. balantidiasis

Balantidium coli.

1455. blastomycosis

Blastomyces dermatitidis.

1456. Bolivian hemorrhagic fever

Arenavirus.

1457. botulism

Clostridium botulinum.

1458. bronchitis

Parainfluenza virus.

1459. bubonic plague

Yersinia pestis.

1460. Burkitt's lymphoma

Epstein-Barr virus.

1461. candidiasis

Candida albicans.

1462. Chagas' disease

Trypanosoma cruzi.

1463. chancroid

Haemophilus ducreyi.

1464. chickenpox

Varicella-zoster virus.

1465. chigger dermatitis

Trombicula mite.

1466. chigger infestation

Tunga penetrans (sandflea).

1467. Chinese liver fluke

Clonorchis sinensis.

1468. cholera

Vibrio cholerae.

1469. coccicioidomycosis (valley fever)

Coccidioides immitis.

1470. colds

Rhinovirus, coronavirus.

1471. condyloma (genital wart)

Human papilloma virus.

1472. conjunctivitis

Haemophilus aegyptius.

1473. crab lice

Phthirus pubis.

1474. cryptococcosis

Cryptococcus neoformans.

1475. **cryptosporidiosis**

Cryptosporidium sp.

1476. **cytomegalic inclusion disease**

Cytomegalovirus.

1477. **Dengue fever**

Dengue virus.

1478. **diphtheria**

Corynebacterium diphtheriae.

1479. **elephantiasis**

Wuchereria bancrofti.

1480. **encephalitis**

Bunyavirus; Eastern, Venezuelan, Western equine encephalitis viruses; St. Louis encephalitis virus.

1481. **endemic (murine) typhus**

Rickettsia typhi.

1482. **epidemic typhus**

Rickettsia prowazekii.

1483. **ergotism**

Claviceps purpura.

1484. **fasciolopsiasis**

Fasciolopsis buski.

1485. **food poisoning (most common bacteria)**

Staphylococcus aureus, Streptococcus pyogenes, Clostridium perfringens, Clostridium botulinum.

1486. food poisoning (other)

Bacillus cereus, Listeria monocytogenes, Campylobacter sp., Shigella sp., Salmonella sp., Vibrio parahemolyticus.

1487. gas gangrene

Clostridium perfringens and other species.

1488. genital herpes

Usually herpes simplex type 2 virus.

1489. giardiasis

Giardia intestinalis.

1490. gonorrhea

Neisseria gonorrhoeae.

1491. granuloma inguinale

Calymmatobacterium granulomatis.

1492. Guinea worm

Dracunculus medinensis.

1493. Hansen's disease (leprosy)

Mycobacterium leprae.

1494. heart worm disease

Dirofilaria immitis.

1495. hemorrhagic fever

Ebola and Marburg filoviruses.

1496. histoplasmosis

Histoplasma capsulatum.

1497. hookworm

Ancylostoma duodenale, Necator americanus.

1498. infectious hepatitis

Hepatitis A virus.

1499. infectious mononucleosis

Epstein-Barr virus.

1500. influenza

Influenza viruses.

1501. kala azar

Leishmania donovani.

1502. keratitis

Acanthamoeba culbertsoni

1503. keratoconjunctivitis

Adenoviruses.

1504. Korean hemorrhagic fever

Bunya virus (Hantaan).

1505. Lassa fever

Arenavirus.

1506. Legionnaire's dissease

Legionella pneumophila.

1507. leishmaniasis

Leishmania braziliensis.

1508. **leptospirosis**

Leptospira interrogans.

1509. **listeriosis**

Listeria monocytogenes.

1510. **liver/lung fluke**

Paragonimus westermani.

1511. **loaiasis**

Loa loa.

1512. **Lyme disease**

Borrelia burgdorferi.

1513. **lymphogranuloma venereum**

Chalmydia trachomatis.

1514. **madura foot**

Actinomadura, Streptomyces, Nocardia sp.

1515. **malaria**

Plasmodium sp.

1516. **measles (rubeola)**

Measles virus.

1517. **meningitis**

Haemophilus influenzae, Neisseria meningitidis, Streptococcus pneumoniae, Listeria monocytogenes.

1518. **meningoencephalitis**

Herpes virus.

1519. molluscum contagiosum

Pox virus group.

1520. mumps

Paramyxovirus.

1521. muscle damage

Coxsackie virus.

1522. non-A, non-B hepatitis

Presumed hepatitis viruses.

1523. oral herpes

Usually herpes simplex type 1 virus.

1524. oriental sore

Leishmania tropica.

1525. ornithosis (psittacosis)

Chlamydia psittaci.

1526. Oroyo fever (Carrion's disease)

Bartonella bacilliformis.

1527. papilloma (common wart)

Human papilloma virus.

1528. pediculosis

Pediculus humanus.

1529. pharyngitis

Streptococcus pyogenes.

1530.   pinworm

Enterobius vermicularis.

1531.   pneumocystic pneumonia

Pneumocystis carinii.

1532.   pneumonia

Streptococcus pneumoniae, Klebsiella pneumoniae.

1533.   pneumonia (also)

Respiratory suncytial virus.

1534.   pneumonic plague

Yersinia pestis.

1535.   poliomyelitis

Poliovirus.

1536.   pseudomembranous colitis

Clostridium difficile.

1537.   puerperal fever

Streptococcus pyogenes.

1538.   Q fever

Coxiella burnetii.

1539.   rabies

Rabies virus.

1540.   rat bite fever

Spirillum minus, Streptobacillus moniliformis.

1541. recrudescent (Brill-Zinnser) typhus

Rickettsia prowazekii.

1542. relapsing fever

Borrelia recurrentis.

1543. respiratory infection

Adenovirus, polyomavirus.

1544. rhinitis

Parainfluenza virus.

1545. rickettsial pox

Rickettsia akari.

1546. Rift valley fever

Bunyavirus (phlebovirus).

1547. river blindness

Onchocerca volvulus.

1548. Rocky Mountain spotted fever

Rickettsia rickettsii.

1549. rubella (German measles)

Rubellavirus.

1550. salmonellosis

Salmonella sp.

1551. scabies

Scarcoptes scabiei.

1552. schistosomiasis

Schostosoma sp.

1553. scrub typhus (tsutsugamushi disease)

Rickettsia tsutsugamushi.

1554. serum hepatitis

Hepatitis B virus.

1555. sheep liver fluke

Fasciola hepatica.

1556. shigellosis (bacillary dysentery)

Shigella sp.

1557. shingles

Varicella-zoster virus.

1558. sickle cell aplastic crisis

Parvovirus.

1559. skin and wound infections

Staphylococcus aureus, S. epidermidis, Streptococcus sp., Escherichia coli, Providencia, Pseudomonas, Serratia.

1560. sleeping sickness

Trypanosoma gambiense, T. rhodesiense.

1561. smallpox

Variola (major and minor) viruses.

1562. sporotrichosis

Sporothrix schenckii.

1563. stongyloidiasis

Strongyloides stercoralis.

1564. swimmer's itch

Schistosoma sp.

1565. syphilis

Treponema pallidum.

1566. tapeworm infestation

Hymenolepsis nana, Taenia saginata, T. solium, Diphyllo-
bothrium latum, Echinococcus granulosis.

1567. tetanus

Clostridum tetani.

1568. tinea (ringworm)

Epiderophyton, Trichophyton, and Microsporium sp.

1569. toxic shock syndrome

Staphylococcus aureus.

1570. toxoplasmosis

Toxoplasma gondii.

1571. trachoma

Chlamydia trachomatis.

1572. trench fever

Rochalimaea quintana.

1573. trichinosis

Trichinella spiralis.

1574. trichomoniasis

Trichomonas vaginalis.

1575. tuberculosis

Mycobacterium tuberculosis.

1576. tularemia

Francisella tularensis.

1577. typhoid fever

Salmonella typhi.

1578. undulant fever (brucellosis, Malta fever)

Brucella sp.

1579. verruga peruana

Bartonella bacilliformis.

1580. visceral larva migrans

Toxcara sp.

1581. whipworm

Tricuruis trichiura.

1582. whooping cough

Bordatella pertussis.

1583. yellow fever

Yellow fever virus.

1584. yersiniosis

Yersinia enterocolitica.

1585. zygomyocosis

Rhizopus sp., Mucor sp.

Terms About Diseases

1586. **abscess**

Accumulation of pus in damaged tissue.

1587. **aplastic crisis**

Period during which erythrocytes are not produced.

1588. **bacteremia**

Presence of bacteria in the blood but without multiplying there.

1589. **bubo**

Enlarged, pus-filled lymph node, especially in plague.

1590. **carbuncle**

Massive pus-filled lesion usually on back or neck.

1591. **casseous lesion**

A lesion with the appearance of cheese curd.

1592. **catarrhal stage**

Stage of a disease characterized by inflammation of mucous membranes and cough.

1593. **catheter**

Hollow tube used in a variety of medical procedures.

1594. **cellulitis**

Diffuse spreading infection of subcutaneous tissue.

1595. **cold sore**

Lesion produced by a herpes simplex virus; fever blister.

1596. consolidation

Blockage of air spaces by fibrin deposits in lobar pneumonia.

1597. coryza

Nasal mucus.

1598. cyanosis

Bluishness of skin due to poor oxygenation of blood.

1599. debridement

Scraping of eschar from a burn to reach infection sites.

1600. dysentery

Severe diarrhea often with blood and mucus in stool.

1601. dysuria

Burning on urination.

1602. enteritis

Inflammation of the digestive tract.

1603. eschar

Thick scab that forms over severely burned tissue.

1604. exanthema

A skin rash.

1605. exudation

Release of soluble substances by living organisms, such as plant roots.

1606. fever blister

Cold sore.

1607. folliculitis

Infection of a hair follicle; pimple or pustule.

1608. fulminating

Displaying sudden, severe symptoms.

1609. furuncle

Deep, pus-filled infection.

1610. gumma

A granulomatous inflammation.

1611. hepatitis

Liver inflammation.

1612. intubation

Introduction of a tube into a hollow organ.

1613. ischemia

Reduced blood flow, which allows wastes to accumulate and deprives cells of nutrients and oxygen.

1614. Koplik spot

Lesion with bluish-white center found in the oral cavity in measles (rubeola) patients.

1615. miliary

Having many very small lesions resembling millet seed.

1616. needle aspiration

Withdrawal of blood or another fluid into a tube.

1617. Negri body

Aggregation of viruses or viral components in brain neuron of rabies infected animal.

1618. nocturia

Need to urinate during the night.

1619. petechia

Tiny red spots in skin folds indicative of hemorrhages seen especially in rickettsial diseases.

1620. pleurisy

Inflammation of pleura.

1621. pus

Fluid containing dead phagocytes, microorganisms, and tissue debris.

1622. Reye's syndrome

Acute disease involving brain edema and liver dysfunction associated with previous viral infection.

1623. sputum

Mucous secretion from the respiratory system.

1624. stridor

Noisy, high-pitched breathing.

1625. sty

An infection of a follicle around an eyelash.

1626. subacute sclerosing panencephalitis

Nearly always fatal persistence of measles virus in brain tissue following a measles infection.

1627. swarmer cell

Spherical, flagellated cell of Rhizobium that forms nodules in roots of legumes.

1628. tetany

A continuous muscle spasm.

1629. tubercle

Solid lesion or granuloma in lungs of tuberculosis patient.

1630. vasculitis

Blood vessel inflammation.

1631. whitlow

A lesion on a finger caused by herpes viruses.

Terms About Infectious Organisms

1632. **choleragen**

Enterotoxin released by Vibrio cholerae after it is ingested.

1633. **Dane particle**

The antigenic portion of the hepatitis B virus.

1634. **Delta agent**

A defective RNA virus that can cause disease only in the presence of a hepatitis B virus infection.

1635. **dental plaque**

Film of bacteria and their secretions on teeth.

1636. **diphtheroid**

Organism that looks like causative agent for diphtheria but fails to produce exotoxin.

1637. **enteroinvasive strain**

A strain of E. coli with a plasmid coding for a surface antigen (K antigen) that allows invasion of mucosal cells.

1638. **exfoliatin**

An exotoxin from Staphylococcus aureus that causes scalded skin syndrome and associated separation of skin layers.

1639. **fluoride**

A chemical element that reduces tooth decay by poisoning bacterial enzymes and hardening tooth enamel.

1640. **food infection**

Gastrointestinal illness caused by ingesting microorganisms that grow and produce toxins in the digestive tract.

1641. **food intoxication**

Gastrointestinal illness caused by ingesting toxins produced in food before it was ingested.

1642. leproma

Large, disfiguring skin lesion in lepromatous Hansen's disease.

1643. lepromin

Agent detected in a diagnostic test for Hansen's disease.

1644. leptomin test

Diagnostic test for Hansen's disease (leprosy).

1645. methacrylate

Substance used to seal tooth surfaces against decay.

1646. murine

Pertaining to mice and rats.

1647. pannus

Superficial growth of blood vessels over the cornea accompanied by growth of granulation tissue.

1648. phagovar

A specific type of phage.

1649. plasmid fingerprinting

Procedure for matching plasmids to identify microorganisms as belonging to the same strain.

1650. retrofection

Reinfection by reentry of agent to body from which it came, as pinworm larvae hatching on buttocks and entering mouth.

1651. serovar

A strain or subspecies.

1652. tetanolysin

A hemolysin produced by Clostridium tetani that contributes to tissue destruction.

1653.   tetanospasmin

A neurotoxin produced by a plasmid gene that causes muscle spasms of tetanus.

1654.   vegetation

A bacterial growth on damaged surfaces of a heart valve.

Environmental Microbiology

1655. **abiotic**

Without life.

1656. **actinorrhizal**

A kind of association between an actinomycete and a plant root.

1657. **activated sludge system**

A sewage treatment involving aerobic organisms and agitation of effluent from primary treatment.

1658. **allochthonous**

Pertaining to organisms introduced to a habitat.

1659. **anaerobic digestion**

Microbial decomposition of sewage in anaerobic conditions.

1660. **arbuscule**

Branching structures formed by endotrophic mycorrhizal fungi in plant roots.

1661. **assimilatory reduction**

Chemical reduction and incorporation of inorganic matter into organic matter of an organism.

1662. **associative nitrogen fixation**

Bacterial nitrogen fixation in a plant root zone.

1663. **autochthonous**

Pertaining to microorganisms or substances indigenous to an environment.

1664. **bacteroid**

A modified bacterium living in and fixing nitrogen in root nodule cells of legumes.

1665. **biochemical oxygen demand (BOD)**

Oxygen needed to maintain organisms in an environment such as a pond or stream.

1666. **biocontrol**

Use of one organism or its products to control another.

1667. **biogeochemical cycles**

Systems in which nutrients and water are recycled.

1668. **biomagnification**

Increase in the concentration of a substance as it goes up a food chain.

1669. **bioremediation**

Use of a process in an organism to counteract a detrimental change such as pollution.

1670. **biotic**

Pertaining to something living.

1671. **bulking sludge**

Sludge that fails to settle during sewage treatment, usually because of growth of filamentous microorganisms.

1672. **carbon cycle**

Process in which atmospheric carbon dioxide enters and is recycled through living and nonliving things.

1673. **chemical oxygen demand (COD)**

Amount of chemical oxidation needed to degrade organic matter in water to carbon dioxide.

1674. **coliform**

Gram-negative, facultative rods that ferment lactose.

1675. **community**

The array of organisms within a given environment.

1676. **confirmed test**

Second stage of coliform testing in which samples from highest dilutions producing gas are streaked on EMB agar.

1677. **consumers**

Organisms that eat producers in order to obtain nutrients.

1678. **decomposers**

Organisms that digest the wastes or dead bodies of producers and consumers.

1679. **decompser-reducer**

Organism that decomposes and reduces the size of a particle.

1680. **dissimilatory reduction**

Use of a substance as an electron acceptor.

1681. **diurnal oxygen shift**

Change in oxygen in water with day night changes in photo-synthesis and respiration.

1682. **ecology**

The branch of biology dealing with the relationships among organisms and their environment.

1683. **ecosystem**

The biotic and abiotic components of an environment.

1684. **ectendomycorrhizal**

Mutualistic association in which fungi surround and partly penetrate plant roots.

1685. **ectomycorrhizal**

Mutualistic association in which fungi surround a root tip with a sheath.

1686. **endogenous respiration**

Use of stored nutrients to maintain an organism without growth.

1687. endomycorrhizal

A mutualistic association in which a fungus penetrates plant root cells and forms arbuscules and vesicles.

1688. endophyte

A plant that lives within another plant.

1689. epilimnion

A warm surface layer of water in a thermally stratified lake.

1690. extended aeration

Sewage aeration sufficient to allow digestion of waste digesting microorganisms themselves.

1691. fecal coliform test

A test for coliforms that normally inhabit the intestinal tract.

1692. flocculation

Precipitation of suspended colloids by the addition of alum.

1693. food web

A pattern of interconnected food chains.

1694. fungistasis

Inhibition of fungal growth.

1695. groundwater

Water in totally saturated ground, which supplies wells and springs.

1696. heterotrophic nitrification

Nitrification done by chemoheterotrophic organisms.

1697. humus

Organic, nonliving soil components.

1698. **hydrologic cycle**

System in which water is recycled, through organisms and by precipitation and evaporation; water cycle.

1699. **hypolimnion**

Deep part of lake with low oxygen and cold temperature.

1700. **indicator organism**

An organism whose presence confirms some condition in the environment, as E. coli indicates fecal contamination.

1701. **indigenous**

Native to an environment.

1702. **infection thread**

Tube formed as nitrogen-fixing bacteria infect a root.

1703. **Liebig's law of the minimum**

Organisms and populations grow until some factor limits their growth.

1704. **membrane filter technique**

Use of a membrane filter to collect microorganisms from air, food, or water.

1705. **microbiostasis**

Inhibition of microbial growth.

1706. **microcosm**

A small part of nature separated for study.

1707. **microenvironment**

Immediate environment around an organism.

1708. **multiple-tube fermentation**

Three-step procedure for detecting coliform bacteria in drinking water.

1709. natural environment

Association of primary producers and decomposers that support consumers.

1710. nitrogen cycle

System whereby nitrogen moves from the atmosphere through certain organisms and back into the atmosphere.

1711. nitrogen fixing bacteria

Bacteria that reduce atmospheric nitrogen to ammonia.

1712. nitrogen oxygen demand (NOD)

Oxygen required by nitrifying organisms in a sewage treatment facility.

1713. nonindigenous

Temporarily, but not permanently, found in an environment.

1714. oligotrophic

Pertaining to a low nutrient environment or organisms that survive in such an environment.

1715. phosphorus cycle

Process in which phosphorus moves between organic and inorganic forms.

1716. population

All the members of a species in a given area.

1717. potable

Safe to drink.

1718. presence-absence test

A simple test for coliforms using lactose broth.

1719. presumptive test

First stage in multiple-tube fermentation in which gas production in lactose broth gives evidence of coliforms.

1720. **primary treatment**

The removal of particulate matter as a first step in sewage treatment.

1721. **producer**

Organism that captures energy and fixes carbon.

1722. **rapid sand filter**

A bed of sand used to trap impurities and to purify water passed through it.

1723. **recalcitrance**

Resistance of a substance (to microbial action).

1724. **rhizosphere**

A region around a plant root where materials from the root stimulate microbial growth.

1725. **root nodule**

Round body on legume roots that contains nitrogen fixing bacteria.

1726. **secondary treatment**

Use of microorganisms to degrade solid wastes after primary treatment of sewage.

1727. **sedimentation basin**

Holding basin for allowing solid wastes to settle out of raw sewage.

1728. **septic tank**

Device used to hold small amounts of domestic sewage while degraded by microorganisms.

1729. **settling basin**

Container used to precipitate solids from sewage by coagulation or flocculation.

1730. **sewage treatment**

Procedure for rendering sewage harmless.

1731. **slow sand filter**

A bed of sand through which water passes slowly because of a film of gelatinous microorganisms.

1732. **sludge**

Solid matter from waste treatment containing aerobes able to digest organic matter.

1733. **sludge digester**

Fermentation tank in which anaerobic bacateria decompose solid waste to simple organic molecules.

1734. **soil pathogen**

Disease causing organism found in soil.

1735. **sulfate reduction**

Use of sulfate to oxidize another substance; addition of sulfur to organic molecules.

1736. **sulfur cycle**

Cycling of sulfur between living and nonliving components of an ecosystem.

1737. **sulfur oxidation**

Oxidation of sulfur to sulfate.

1738. **sulfur reduction**

Reduction of sulfur to hydrogen sulfide.

1739. **tertiary treatment**

Biological removal of materials such as metals, minerals, and viruses from the efflux of secondary sewage treatment.

1740. **thermocline**

The layer in a thermally stratified lake that separates the epilimnion from the the hypolimnion.

1741. **total organic carbon (TOC)**

Total carbon in organic molecules within a biological sample.

1742. trickling filter system

Use of a bed of rocks coated with aerobic organisms for decomposing organic matter in sewage as it "trickles" in.

1743. tripartite association

An association of a plant and two kinds of fungi.

1744. VA mycorrhiza

An endomycorrhiza having vesicles and arbuscules found in plant roots.

1745. vesicle

Small membrane bound body.

1746. water cycle

System in which water is recycled, through organisms and by the processes of precipitation and evaporation.

1747. Winogradsky column

Glass column with anaerobic to aerobic zones simulating microbial growth conditions in a nutrient-rich lake.

Applied Microbiology

1748. **aflatoxin**

Carcinogenic fungal toxin.

1749. **biochemical marker**

An inherited biochemical characteristic monitored to detect genetic change.

1750. **biocide**

An agent that can kill living organisms.

1751. **bioconversion**

Use of organisms to make a product the organisms do not normally need for growth.

1752. **biodegradation enhancement**

Environmental changes that foster breakdown of certain substances.

1753. **biodeterioration**

Detrimental change caused by organisms, especially microorganisms.

1754. **biopolymer**

A large molecule with repeating units made by an organism.

1755. **biosensor**

A process in an organism used to detect a particular substance.

1756. **biotransformation**

Cnversion of one substance to another by an organism.

1757. **bottom yeast**

In beer production, yeast that typically settles to the bottom of a fermentation vat.

1758. canning

Preservation of food by heating sealed cans to destroy pathogens.

1759. continuous feed

Continuous addition of nutrients to a culture.

1760. idiophase

Period in a culture when secondary rather than primary metabolites are synthesized.

1761. industrial microbiology

Branch of microbiology that produces useful products or services.

1762. lagered

Pertaining to the aging of beer.

1763. malt

Grain soaked in water to start germination and enzyme action in preparation for brewing and distilling.

1764. mash

Soluble materials from germinated grain used as microbial growth medium.

1765. mashing

Process by which cereals are incubated to degrade complex carbohydrates to sugars.

1766. microbial transformation

Use of microorganisms to make substances not needed for their own growth; bioconversion.

1767. must

Juices that can be fermented to alcohol.

1768. non-Newtonian broth

A plastic-like liquid that changes in viscosity with rate of agitation.

1769. osmophilic

Growing best in a medium with a high osmotic concentration.

1770. pharmaceutical microbiology

Application of microbiology to the development of drugs and similar products.

1771. pitched

Pertaining to inoculation of a medium with yeast as in beer brewing.

1772. primary metabolite

A microbial metabolite produced during a growth phase.

1773. putrefaction

Microbial decomposition of organic matter in which foul smelling substances are released.

1774. racking

Removal of sediment from wine.

1775. radappertization

Use of gamma rays from cobalt to kill microorganisms in food.

1776. regulatory mutant

Mutated regulator gene.

1777. scaleup

Extrapolation of small scale activities to large scale activities, as in industrial applications.

1778. secondary metabolite

Metabolic products synthesized after growth is complete.

1779. single-cell protein (SCP)

Food protein extracted from single-cell organisms.

1780. **sour mash**

Mash inoculated with lactic-acid producing organisms to control undesirable growth and achieve desirable taste.

1781. **starter culture**

An inoculation of carefully selected organisms to start a commercial fermentation.

1782. **trophophase**

Active growth phase in a batch culture in which primary (growth-directed) metabolism is dominant.

1783. **wine vinegar**

Vinegar made by oxidation of alcohol in wine to acetic acid.

1784. **wort**

Filtrate from malted grain that is fermented in beer making.

Index

www.ingramcontent.com/pod-product-compliance
Lightning Source LLC
Chambersburg PA
CBHW081119170526
45165CB00008B/2487